Demystifying OWL for the Enterprise

Synthesis Lectures on the Semantic Web: Theory and Technology

Editors
Ying Ding, *Indiana University*
Paul Groth, *Elsevier Labs*
Founding Editor
James Hendler, *Rensselaer Polytechnic Institute*

Synthesis Lectures on the Semantic Web: Theory and Technology is edited by Ying Ding of Indiana University and Paul Groth of Elsevier Labs. Whether you call it the Semantic Web, Linked Data, or Web 3.0, a new generation of Web technologies is offering major advances in the evolution of the World Wide Web. As the first generation of this technology transitions out of the laboratory, new research is exploring how the growing Web of Data will change our world. While topics such as ontology-building and logics remain vital, new areas such as the use of semantics in Web search, the linking and use of open data on the Web, and future applications that will be supported by these technologies are becoming important research areas in their own right. Whether they be scientists, engineers or practitioners, Web users increasingly need to understand not just the new technologies of the Semantic Web, but to understand the principles by which those technologies work, and the best practices for assembling systems that integrate the different languages, resources, and function-alities that will be important in keeping the Web the rapidly expanding, and constantly changing, information space that has changed our lives.

Topics to be included:

- Semantic Web Principles from linked-data to ontology design

- Key Semantic Web technologies and algorithms

- Semantic Search and language technologies

- The Emerging "Web of Data" and its use in industry, government and university ap-plications

- Trust, Social networking and collaboration technologies for the Semantic Web

- The economics of Semantic Web application adoption and use

- Publishing and Science on the Semantic Web

- Semantic Web in health care and life sciences

Demystifying OWL for the Enterprise
Michael Uschold
2018

Validating RDF
Jose Emilio Labra Gayo, Eric Prud'hommeaux, Iovka Boneva, and Dimitris Kontokostas
2017

Natural Language Processing for the Semantic Web
Diana Maynard, Kalina Bontcheva, and Isabelle Augenstein
2016

The Epistemology of Intelligent Semantic Web Systems
Mathieu d'Aquin and Enrico Motta
2016

Entity Resolution in the Web of Data
Vassilis Christophides, Vasilis Efthymiou, and Kostas Stefanidis
2015

Library Linked Data in the Cloud: OCLC's Experiments with New Models of Resource Description
Carol Jean Godby, Shenghui Wang, and Jeffrey K. Mixter
2015

Semantic Mining of Social Networks
Jie Tang and Juanzi Li
2015

Social Semantic Web Mining
Tope Omitola, Sebastián A. Ríos, and John G. Breslin
2015

Semantic Breakthrough in Drug Discovery
Bin Chen, Huijun Wang, Ying Ding, and David Wild
2014

Semantics in Mobile Sensing
Zhixian Yan and Dipanjan Chakraborty
2014

Demystifying OWL for the Enterprise
Michael Uschold

ISBN: 978-3-031-79481-0 print
ISBN: 978-3-031-79482-7 ebook

DOI 10.1007/978-3-031-79482-7

A Publication in the Springer series
SYNTHESIS LECTURES ON THE SEMANTIC WEB: THEORY AND TECHNOLOGY #17

Series Editors: Ying Ding, Indiana University, Paul Groth, Elsevier Labs
Founding Editor: James Hendler

Series ISSN: 2160-4711 Print   2160-472X Electronic

Demystifying OWL for the Enterprise

Michael Uschold

Semantic Arts, Inc.

SYNTHESIS LECTURES ON THE SEMANTIC WEB: THEORY AND TECHNOLOGY #17

ABSTRACT

After a slow incubation period of nearly 15 years, a large and growing number of organizations now have one or more projects using the Semantic Web stack of technologies. The Web Ontology Language (OWL) is an essential ingredient in this stack, and the need for ontologists is increasing faster than the number and variety of available resources for learning OWL. This is especially true for the primary target audience for this book: modelers who want to build OWL ontologies for practical use in enterprise and government settings. The purpose of this book is to speed up the process of learning and mastering OWL. To that end, the focus is on the 30% of OWL that gets used 90% of the time.

Others who may benefit from this book include technically oriented managers, semantic technology developers, undergraduate and post-graduate students, and finally, instructors looking for new ways to explain OWL.

The book unfolds in a spiral manner, starting with the core ideas. Each subsequent cycle reinforces and expands on what has been learned in prior cycles and introduces new related ideas.

Part 1 is a cook's tour of ontology and OWL, giving an informal overview of what things need to be said to build an ontology, followed by a detailed look at how to say them in OWL. This is illustrated using a healthcare example. Part 1 concludes with an explanation of some foundational ideas about meaning and semantics to prepare the reader for subsequent chapters.

Part 2 goes into depth on properties and classes, which are the core of OWL. There are detailed descriptions of the main constructs that you are likely to need in every day modeling, including what inferences are sanctioned. Each is illustrated with real-world examples.

Part 3 explains and illustrates how to put OWL into practice, using examples in healthcare, collateral, and financial transactions. A small ontology is described for each, along with some key inferences. Key limitations of OWL are identified, along with possible workarounds. The final chapter gives a variety of practical tips and guidelines to send the reader on their way.

KEYWORDS

OWL, ontology engineering, data modeling, conceptual modeling, Semantic Web, knowledge graph, enterprise ontology, semantic technology, semantics, reuse, modularity, metadata, resource description framework (RDF), RDF Schema, triples, description logic, knowledge representation

Contents

Foreword by Dave McComb

THE WORLD NEEDS THIS BOOK

The Semantic Web launched in 2001 with Tim Berners-Lee's article in *Scientific American*. By 2004, the W3C had finalized the standardization of the OWL language for modeling ontologies. OWL and the related standards RDF and RDFS enjoyed a brief period of interest, perhaps even hype, in 2007–2009. The interest was short lived.

There were two related reasons that early adopters abandoned the Semantic Web: (1) it was perceived as being too complicated and (2) practitioners didn't understand it.

The "too complicated" rap was partly due to the fact that it is a fairly complex spec, but it was reinforced by the many books, articles, and tutorials that came out at the time. It almost seemed as if the authors intentionally wanted to encourage the perception of a high priesthood, only able to be fathomed by the chosen few.

The "didn't understand it" rap was partly due to the complexity, but persisted because developers and modelers tried to recreate the style of model they were comfortable with. Developers build object-oriented-looking ontologies, and relational database modelers built ontologies that looked a lot like the ER models they were familiar with. Each group was disappointed when they couldn't implement simple constraints, and were frustrated when they finally came up against the "open world assumption."

This is unfortunate because what we have found in the intervening decade is that this modeling language and the technologies that come along with it are the best bet for reversing the siloed mess that most large enterprises deal with on a daily basis.

What Michael has done here is to create a shallow end of this swimming pool. This is a gentle introduction that anyone, even those with the least background in technology or modeling, can easily follow. He introduces and thoroughly explains the 30% of the OWL spec that practicing ontologists use on an everyday basis. He does this without glossing over the things that make OWL and Semantics so special.

Coming away from this book, you will understand the special place that OWL should have in your enterprise datascape. You will understand how OWL fits in with other standards and technologies. You will appreciate how it can simplify your enterprise ontology like no other technology can.

Michael is uniquely suited for the task of writing this book. His Ph.D. was in Artificial Intelligence, specializing in ontology-driven development of ecological simulation software. He was a very early builder and user of ontologies.

After leaving academia he worked in industry designing and applying ontologies for Boeing and Reinvent. He joined Semantic Arts in 2010. In that time he has designed over ten enterprise ontologies and has taught hundreds of budding ontologists in the subtleties of this technology.

This combination of theoretical background coupled with pragmatic experience is, to the best of my knowledge, unequaled.

It has been a pleasure working with Michael at Semantic Arts these seven-plus years.

Enjoy this book, and welcome to the next generation of enterprise information systems!

Dave McComb

Foreword by Mark A. Musen

ONTOLOGIES ARE EVERYWHERE

We can't order merchandize online, stream a movie, search the Web, or access social media without interacting with software that uses ontologies. Most of the software that surrounds us and that we often take for granted has at its core ontologies—almost always taken for granted—that characterize the merchandize, the movies, the websites, the users, and everything else that the software needs to compute about. Whether they are actually called ontologies, or product catalogs, or knowledge graphs, data structures that capture models of the entities in the world with which a system interacts are critical components of modern computing technologies.

But ontologies sure are hard to build. Understanding how the entities in some application area might be modeled, how they might interrelate with one another, and how they might be captured in software is really difficult. The problem is exacerbated because the standard computer language for representing ontologies—OWL—is complex and often nonintuitive. OWL causes all kinds of problems for new ontology engineers. In OWL, it's easy to infer that left is right and that up is down, unless you are extremely careful. An OWL ontology that states that an opera is a play in which all the words are sung will also classify as an opera a pantomime—a play that has no words at all! If you're not vigilant, OWL might tell you that a gall bladder is a golf club or that a toothbrush is a sonnet.

Why do we put up with this nonsense? We use OWL because it has many useful properties that allow us to understand the implications of our modeling choices, enabling us to have more confidence that we have modeled things correctly. We also use OWL because it has become an international standard. Before OWL became a recommendation of the World Wide Web Consortium, there was no prevailing language for encoding ontologies, no easy way to integrate ontologies, and few widely used tools for building ontologies. That chaos disappeared with the advent of OWL. Standardization simplified many practical aspects of ontology engineering and allowed the ontology-development community to share ontology content, ontology-engineering systems, and best practices for ontology engineering on a broad scale.

The Protégé ontology editor developed by my group at Stanford University was the first widely used tool for building ontologies in OWL, and it remains the only open-source platform that supports OWL-based ontology development that is in common use. Over the years, we've continued to enhance Protégé with additional features that, we believe, help users to deal with many of the complexities of ontology engineering in OWL. But none of these features overcomes the basic problem that description logics in general, and OWL in particular, have elements that are unintuitive and hard to learn.

I teach about OWL at Stanford, and students who are new to the language always seem to end up building ontologies that classify toothbrushes as sonnets, scratching their heads trying to figure out why. OWL's "open world assumption" and its somewhat arcane methods for defining the characteristics of the entities in a model are initially hard to grasp. Students are confused that language developers would choose to make things so difficult and complicated. They persevere, however, because they appreciate the importance of ontology engineering in the development of many modern software systems and the critical role that ontologies play in many modern professional activities, particularly in the sciences. Students still struggle, both because the syntax is difficult and because the implications of their modeling choices are often hard to predict. As an instructor, I have been frustrated that there is no easy way to teach students the basic components of knowledge modeling in OWL other than to sit them down in front of an ontology editor such as Protégé and ask them to represent the essential features of a toothbrush or a golf club. To date, all the writings about OWL have come from scientists in the description-logic community whose main objectives have been completeness and accuracy rather than pedagogy and understanding. The literature is thus one written by experts for experts, leaving novice learners to stare helplessly at pages of complex equations and at unpronounceable abbreviations such as SHOIN and SHROIQ, always set in ridiculous fonts.

This book is different. It clearly "demystifies" OWL by distilling the language to its very basic features and by presenting clear, easy-to-follow examples. The emphasis is on communicating clearly the core elements of the language, rather than on expansiveness and logical rigor. This volume is an important contribution, coming at an important time, as ontologies enter the mainstream of software engineering and are no longer in the exclusive domain of highly specialized experts.

This book marks an important transition. At last, there is recognition that ontologies are often being constructed in the course of routine software development. It is now apparent that there is a need for more traditional software engineers to be able to create ontologies and to render them in OWL, the representation language that has now entered standard usage. This is a natural evolution in the trend toward translation of human knowledge into computable forms, enabling new technologies to interact with people and to communicate human ideas in novel ways. Most people don't think critically about the ontologies that allow them to find the products that they want online or that suggest the content in which they might be interested on social media; their interactions with these ontologies come naturally and implicitly. Similarly, the software engineers who build these semantically aware systems should not get hung up on the complexities or enigmatic properties of OWL; their work to model and represent ontologies should come just as instinctively.

It's about time that OWL became demystified. The next generation of intelligent software systems depends on it.

Mark A. Musen
Palo Alto, California

Preface

WHY OWL?

The current state of information technology in the modern enterprise has been described as a "Software Wasteland".[1] There are countless silos where each application has its own database and its own database schema with consequent duplication and high costs of integration and change. There are a few root causes.

First, there is no mechanism for breaking up a data schema into modules that can be re-purposed and reused across multiple databases. Monolithic data models and the lack of reusability increase the cost of change. Second, there is no way to uniquely identify data or schema elements globally across databases; this results in high integration costs. Finally, and perhaps most importantly, there are no widely used technologies and practices for representing the meaning of the data and schema elements as they evolve. Conceptual models may exist, but they are rarely kept up to date.

After a slow incubation period of nearly 15 years, the modern enterprise is waking up to the value of the Semantic Web stack of technologies, which has addressed all 3 of the above root causes.

The meaning of the data in a semantic application is defined using the Web Ontology Language (OWL). An OWL ontology is a model that represents the subject matter of the data (e.g., healthcare or electrical products) that will reside in triple-store databases that will be used by one or more related applications. OWL is built on the Resource Description Framework (RDF) which is a knowledge graph language based on triples. The support for globally unique identifiers is baked into RDF and thus OWL.

OWL is an essential ingredient in the semantic technology stack
that continues to grow and evolve.

Leading-edge enterprises are building their own enterprise ontologies and enterprise knowledge graphs.[2,3] The technology stack includes a graph query language analogous to SQL called SPARQL for querying triple stores. More recent additions to the stack include R2RML for converting data from relational databases into triples and SHACL for representing constraints and

---

[1] *Software Wasteland* by D. McComb, https://technicspub.com/software_wasteland/,

[2] Exploiting Linked Data and Knowledge Graphs in Large Organisations http://www.springer.com/us/book/9783319456522.

[3] Linking Enterprise Data. http://www.springer.com/us/book/9781441976642.

other details that are pertinent to specific applications. These are separate important tools for use in conjunction with OWL ontologies, and are beyond the scope of this book.

Finally, there are ample industry-scale tools provided by vendors that support these standards. The time is now to step out of the software wasteland.

WHY THIS BOOK?

It's not easy to learn OWL on your own, even if you have a Ph.D. in artificial intelligence and training in formal logic. I found that out the hard way in 2002, when I was tasked with learning a precursor to OWL at a research lab at The Boeing Company. The purpose of this book is to dramatically speed up that learning process for others.

In 2010, I joined Semantic Arts as a senior ontology consultant and have been teaching OWL and using it to build industrial ontologies for the past seven-plus years. As few as five years ago, one of our clients told us not to use the "O" word (ontology)—because it would scare people. Back then, hardly anyone in a typical enterprise knew much about ontology, and there were few if any projects going on. That has dramatically changed in the past five years.

The need for ontologists is growing faster than the number and variety of available resources for learning OWL, especially from an industry perspective. This book differs from others available at the time of writing in that it is focused primarily on the needs of the modeler in an industrial context. I take a demand-pull approach, only introducing an OWL construct when the need for it arises to meet a modeling need. I focus on the 30% of OWL that gets used 90% of the time. Finally, I use examples from real-world industrial ontologies created in my day-to-day work.

The material in the first two chapters of this book has been presented five times in the past four years as a half-day tutorial. Venues included the Semantic Technology Conference, International Semantic Web Conference, Semantic Technology for Intelligence, Defense and Security, Data Architecture Summit, and Enterprise Data World. Some of the material has been folded into the Designing and Building Business Ontology class that I co-teach for Semantic Arts. The success I had with this material inspired me to expand it into a book.

TARGET AUDIENCE

This book is a gentle but thorough introduction to the most important parts of OWL. The only prerequisites for this book are an interest in modeling and a knack for thinking logically. The primary audience consists of modelers who want to build OWL ontologies for practical use in enterprise and government settings. For them, I drive most of the learning from real-world examples and avoid unnecessarily technical language. Secondary target audiences include the following.

- *Intermediate to expert modelers in any setting* having some familiarity with OWL who wish to deepen their understanding and see things from a fresh perspective.

- *Technically oriented managers* who want to know about ontology and OWL to better interact with their technical people.

- *Undergraduates and post-graduates* who want to understand OWL from a practical enterprise perspective.

- *Instructors* who are looking for new ways to explain OWL.

- *Semantic technology developers* who want a better understanding of the OWL ontologies that they code to.

OVERVIEW OF THIS BOOK

The book unfolds in a spiral manner. In the first cycle, I describe the core ideas. Each subsequent cycle reinforces and expands on what has been learned in prior cycles and introduces new related ideas. The book is divided into three parts.

Part 1: Introducing OWL

This is a cook's tour of ontology and OWL, giving an informal overview of what things need to be said to build an ontology, followed by a detailed look at how to say them in OWL. This is illustrated using a healthcare example. I conclude by explaining some foundational ideas about meaning and semantics to prepare for going deeper in the next section.

Part 2: Going into Depth: Properties, Classes, and Inference

Everything to do with building an OWL ontology revolves around properties and classes. I give detailed descriptions of the main constructs that you are likely to need in everyday modeling, including what inferences are sanctioned.

Part 3: Using OWL in Practice

Using examples in healthcare, collateral, and financial transactions, we put into practice what we have learned so far. For each, I describe a model and show some inferences. Next, I identify some key limitations of OWL and possible workarounds. I conclude with a variety of practical tips and guidelines to send you on your way.

STYLE

Per common practice, most of this book is written using the editorial "we." At times, "we" will refer to the collective shared views and experiences of myself and my ontologist colleagues. "I" is used to express a personal perspective that may not be shared by my colleagues.

EXERCISES

In a number of places throughout the book you will find exercises. Answers may be found in Appendix A.4.

Acknowledgments

I first acknowledge those who made this book possible. I am deeply grateful to Dave McComb for providing an opportunity to build so many ontologies in such a wide variety of industries; and for the many insightful ideas he has shared over the years. My views as an ontologist have been greatly influenced by him. Dave also gave insightful feedback on a later draft of this book.

My parents' greatest gift was to encourage me and my eight siblings to pursue *our* own dreams, not theirs. A special thanks to Professor Richard Uschold (my Dad) who shared with me his unique gift for explaining complex technical ideas. Alan Bundy helped me through some tough times in graduate school and taught me how to organize a major writing project.

I have Ying Ding to thank for asking me a couple years ago: "Have you ever thought about writing a book?" I thank Larry Raymore for planting a seed that got me started. I'm grateful to Michael Morgan who agreed to publish this book and has been a pleasure to work with.

My views on OWL itself and how it is most effectively taught have been influenced by the numerous ontologists and modelers that I have worked with over the years. These include Semantic Arts colleagues: Andi Engelstad,[4] Dan Carey, Dave McComb, Mark Ouska, Mark Wallace, Simon Robe, and Ted Hills. It also includes ontologists authoring the Financial Industry Business Ontology (FIBO). These include David Newman, Dean Allemang, Elisa Kendall, Mike Bennet, and Pete Rivett. FIBO was also a source of some of the examples I used in this book. Thanks to Simon Robe also for permission to use his visual OWL syntax.

My hero reviewers were Dalia Varanka and Mark Ouska who read every word, sometimes more than once, and turned around drafts in record time. In addition, I received a lot of useful feedback on different chapters from Andrea Engelstad, Andrea Zachary, Bobbin Teegarden, Dave Plummer, Matt Johnson, and Pete Rivett. Andrea Zachary made a major contribution to the index.

Finally, I am deeply grateful to Mary VonRanker, for listening to and supporting me in numerous ways. She was a very effective sounding board for ideas about the book.

---

[4] All lists of names are in alphabetical order.

Part 1
Introducing OWL

Part 1 introduces the basics of the Web Ontology Language (OWL) using plenty of examples. Building an ontology means creating a model to describe some subject matter of interest in a way that supports automated reasoning. Using OWL consists primarily in being able to represent the following three ideas:

1. individual things, e.g., JaneDoe,

2. kinds of things, e.g., Organization, and

3. kinds of relationships, e.g., worksFor.

We describe the many ways that the above three things can be combined and used. Ontologies and the data that populate them are stored in knowledge graphs composed of triples. Triples assert relationships between things—e.g., JaneDoe worksFor Microsoft. Automated reasoners conclude new information.

Chapter 1 introduces the main ideas using informal conversational terminology that is independent from OWL per se. The focus is to identify the kinds of things that you need to say when you build an ontology. Chapter 2 explains how to say those things in OWL. It introduces the formal OWL terminology for the foundational ideas described in the first chapter. Chapter 3 goes a bit deeper into a variety of things that are important to get going in OWL quickly. It will help you think in the right way and avoid some common mistakes.

By the end of Part 1, you will have seen the 30% of OWL that you will use 90% of the time.

CHAPTER 1

Getting Started:
What Do We Need to Say?

1.1 WHAT IS AN ONTOLOGY? WHAT IS OWL?

There are countless definitions of "ontology" that you may wish to explore further, and whole papers were written on the subject in the formative years of the field. For our purposes, just think of an ontology as *a model which represents some subject matter*. We avoid the common usage of the term "domain" as a synonym for subject matter, because it has a formal meaning in OWL. An ontology communicates what kinds of things there are (for the subject matter of interest) and how they are related to each other. It is built so that automated reasoning software can draw conclusions resulting in new information.

An ontology is different from other models you may have seen in that it represents some (suitably scoped) subject matter as a whole, rather than a model of a particular thing (like an airplane) or a predictive model (of hurricanes, earthquakes, or climate). Also, an ontology can provide a structure for data like a database schema can. However, unlike the latter, an ontology can provide great value even in the absence of data.[1]

Many notations and languages have been used over the years to represent ontologies, some more rigorous than others. The Web Ontology Language (OWL), developed by the World Wide Web Consortium (W3C), is a language for representing ontologies that is based on formal logic, a discipline that evolved from philosophy and mathematics. It is the only standard for representing ontologies that is widely used both in academia and industry. This book uses examples from a variety of industries based on the commercial ontologies that have been developed for our clients.

In the remainder of this chapter, we explore some of the key concepts in a variety of industries that one might wish to model using OWL. We will discover the kinds of things we need to be able to say about the subject matter to be modeled and show many examples. In doing so, we identify requirements for what OWL must be able to express. In Chapter 2, we explain how to express them in OWL.

---

[1] **See:** Ontologies and Database Schema: What's the Difference?

1.2   IN THE BEGINNING THERE ARE THINGS

An ontology is a model that represents some subject matter. For any subject, there are *things* that you care about and want to identify and express in an OWL ontology. What are some specific things in different industries? Say you are a healthcare provider. What are the most important things in healthcare? Stop reading for a moment and brainstorm. Write down a dozen or more of the things that come to mind. For now, focus mostly on things that are written down as nouns (or noun phrases). These nouns will correspond to kinds of things in your subject. Then ask yourself, what are the one, two, or three most fundamental things in healthcare that would concern you as a provider?

Say you are in the finance industry—perhaps a discount broker, or maybe the CIO of a full service investment bank. What are the things that come to mind? What are the most important things in this industry? If it helps, limit the scope to managing assets. Do the same exercise. Brainstorm, write down a dozen or more of the key things, and identify up to three of the most central things in this industry.

Say your job is to manage registration and documentation of ongoing changes for corporations and charities for some particular jurisdiction (e.g., the state of Washington). Go through the same exercise once more. What are the most central things of importance in this subject area?

If none of these examples stimulate you, think of others, ones you know a lot about and have passion for. Pick any subject matter that you would like to have an ontology for. Identify a dozen or so key things, and select the most important ones. In the database field, this activity of identifying what is important in some subject matter is part of data modeling.

Table 1.1: Core ideas for different subjects

Healthcare	Finance	Corporation Registration	Your favorite subject
Disease	Stock	Shares of stock	
Diagnosis	Bond	**Corporation**	
Health insurance policy	Broker	Articles of Incorporation	
Regulations	Asset	Director / CEO	
Medicine	NYSE	Jurisdiction	
Hospital	S&P 500	Statute	
Doctor	Legal Entity	Registered Address	
Nurse	**Trade**	Registered Agent	
Patient	Account balance	Annual report	
Appointment	Fees	Registration fee	
Patient visit	Risk	Official & DBA name	
Credentials	Money	Person	

Table 1.1 shows some of the things you might have written down. The items in bold are central to that subject.

1.3   KINDS OF THINGS VS. INDIVIDUAL THINGS

Whatever the purpose of your ontology, there is always a dance that goes on between thinking about a particular individual vs. the generic kind of thing that it is. For example, Google Inc. is an individual thing. The kind of thing it is, is `Corporation`. We need to be clear whether we are talking about an individual thing, or the kind of thing that individual is.

If John Doe saw the doctor on January 12, 2014, the kind of thing that "seeing the doctor" is, is called a "patient visit." The kind of thing that John Doe is, is patient. If John was treated by Jill Smith, then the kind of thing Jill Smith is, would be, person, and more specifically, a doctor or nurse. The patient, John Doe is also a person, and might also be a cancer patient.

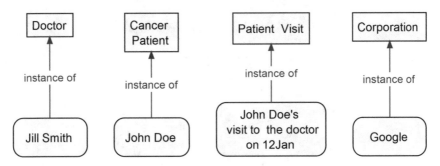

Figure 1.1: Individual things and their kinds. Individuals have rounded corners; the rectangles depict kinds.

Figure 1.1 depicts individuals as shapes with rounded corners and their kinds as rectangles. Pick a few of the concepts listed in Table 1.1, especially those in your favorite subject. Then identify the kinds of things and, for each, identify some specific individuals.

1.4   NO THING IS AN ISLAND

By now, I hope you have written down a few dozen things in a few industries. The things that came to mind likely have important roles to play in that subject area, and thus will be related to a number of other things that are also important for that subject. Decide which of the things you wrote down are most important in the sense that they have the richest set of relationships with other things in the same subject area. These are likely to be the most important things you chose in the last exercise.

1.4.1   HEALTHCARE

Arguably the most central thing in healthcare is the event of a person receiving healthcare of some sort. After all, the primary goal of healthcare is to keep people healthy. Do you have something in your list that includes this? Perhaps a doctor's visit or hospital stay. We will use the more general notion of a patient visit which links to both a patient and a healthcare provider (for example, a doctor). There is also the care that was provided, which often includes a diagnosis and treatment. Figure 1.2 illustrates some key things in healthcare and how they are related.

The dotted lines can be concluded from knowing the solid line connections. If you know who the care recipient and care provider are for a given patient visit, you can conclude who received care from whom (or, conversely, who gave care to whom).

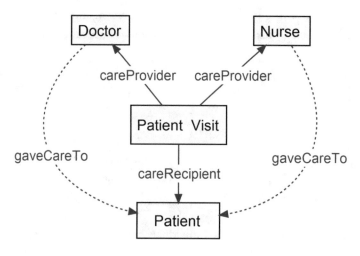

Figure 1.2: Inter-related things in healthcare.

1.4.2   FINANCE

In the asset management side of finance, the goal is to get a return on assets. This is accomplished by a series of financial transactions, most prominently trades. For example, you might purchase 50 shares of Google common stock (GOOG). You might have to sell a bond to free up assets to make that purchase. Each of these transactions is a trade, so the trade is central in asset management.

A trade is not an island. Like a patient visit, it is connected to a variety of other things. What is it related to? There is a buyer, a seller, and possibly a broker. Money changes hands, and ownership is transferred from one legal entity to another along with associated formal documentation. Figure 1.3 illustrates some key things in asset management and how they are related.

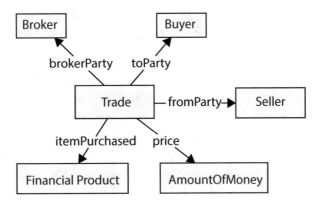

Figure 1.3: Inter-related things in finance.

1.4.3   CORPORATE REGISTRATIONS

Exercise 1: Draw a diagram highlighting the key concepts and relationships for the subject of registering corporations just like we did for finance and healthcare. The idea is to scan your list of items or the list in the third column of Table 1.1 and to pick what you think is the central concept. Then list a few of the key relationships that link it to other concepts that are in the table, or that you think are important but are not in the table.

1.5   THINGS CAN HAVE A VARIETY OF ATTRIBUTES

We can say quite a lot about a given thing by specifying relationships connecting it to other things, as in the above examples. But there is more that we want to say that is not so easily handled in that way. For example, many things have associated names and dates. We may wish to state how old someone is. For example, Jill Smith has the first name, "Jill" and her age is 32. Google's official name is "Google LLC" and it was incorporated on September 4, 1998 (see Figure 1.4).

These kinds of statements are very important, but are different from the other statements we have been making. We have been talking about two individual things being related to one another. But capturing information about names and dates is specifying information about what characteristics or *attributes* a given individual has. Unlike connecting one individual to another, we are connecting something to a literal value, typically a string, a number or a date.

From another perspective, note that it makes sense to say a trade is related to the broker on that trade or that Jane is related to her patients. But it is quite awkward to say "Google is related to the string 'Google LLC'", or that Jane is in a relationship with the string "Smith." Instead of connecting one individual to another, we are connecting an individual to a literal value. Things like "age," "date of incorporation," and "first name" are commonly referred to as *attributes*. Rather than saying one thing is in a relationship with another thing, we say that something has an attribute whose value is a literal of some kind.

The most common kind of literal is a string, which is used for names, descriptions, and many other things. Other kinds of literals include dates and different kinds of numbers like integer or decimal. Numbers will be used for measuring and counting things like weight and age. A date is a specially formatted item with a very specific meaning. From a modeling perspective, the main thing that characterizes a literal is that we won't be saying anything more about particular literals such as "John" or the number 32. Literals can be thought of as pure values; they don't have properties or attributes of their own (see Figure 1.4).

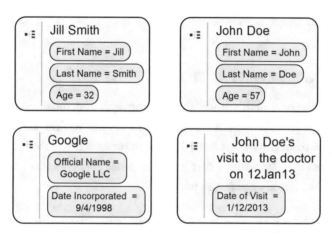

Figure 1.4: Some common attributes and their literal values.

1.6   MORE GENERAL THINGS AND MORE SPECIFIC THINGS

In our examples so far, we have come across different kinds of things, including patient visit, trade, person, doctor, nurse, patient, legal entity, and corporation. Notice that every corporation is a legal entity. Thus, a corporation is a *specific kind* of legal entity. What other kinds of legal entity are there? Persons are legal entities, as are many other kinds of organizations, e.g., partnerships, limited liability companies (LLCs), cooperatives, and some charities to name a few. We can think of a legal entity as generalizing these other kinds of things.

Consider the concepts doctor and nurse. There is a more general kind of thing that each of these can be seen as more specific variations of. What might that be? Both are healthcare providers, and both are also persons. What other examples can you find in the things we have seen so far, where one is more general or more specific than another? Have a look at Figure 1.5 which covers the three subject areas we have considered.

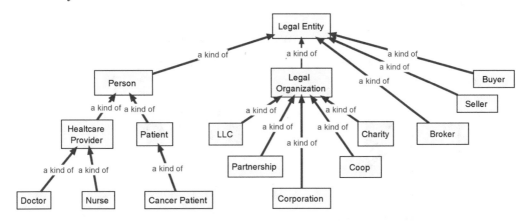

Figure 1.5: A hierarchy of different kinds of things.

This concludes the discussion about what kinds of things you need to say to build an ontology. Recall that we said an ontology was built in such a way as to support drawing conclusions from existing information. We consider that next.

1.7   DRAWING CONCLUSIONS

Despite their remarkable capabilities, computers take things quite literally and at times seem rather dumb. They don't know the most simple and obvious things. Fortunately, that is changing. These days once a man indicates he is male on an online health form, there is good chance it won't bother to ask him whether he is pregnant. This is a simple example of the computer doing something a little smarter. It was able to *draw the conclusion* that the man was not pregnant because it knows that he is male and males cannot be pregnant. Computer programs that are designed to draw conclusions that logically follow from an existing set of data or assertions are called *automated reasoners* or *inference engines*.

While this sort of thing can easily be accomplished through hard-coded rules, that approach does not scale. We want a more general way to tell the computer things and have it apply some general principles that allow it to draw a wide variety of interesting and useful conclusions.

Given our examples so far, can you think of some situations where you would want the computer to automatically draw some conclusions for you? Below are a couple examples to get you started. They are depicted in Figure 1.6. Bold lines are for 'a kind of' links; thinner lines are for 'instance of' links. Solid lines are directly asserted, dotted lines indicate the drawing of conclusions.

1. If a cancer patient is a kind of patient, and a patient is a kind of person, then we want the computer to be able to figure out what common sense tells us, that a cancer patient is also a kind of person.

2. If Google is a corporation, and a corporation is a kind of legal entity, then we want the computer to conclude that Google is in fact a legal entity, just as common sense tells us.

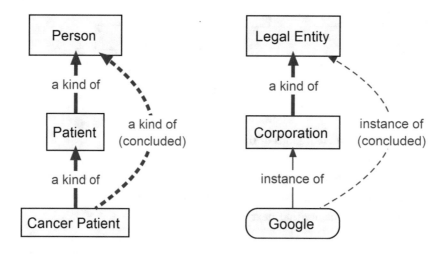

Figure 1.6: Drawing simple conclusions.

When we say the system should be able to "figure out what common sense tells us" we mean it should be able to take the existing information it has and to conclude some new information that *follows logically* from that information. OWL is based on formal logic, which we will discuss in Chapter 3. It specifies exactly when something follows logically from something else and thus what conclusions should be drawn. It's one thing to draw a conclusion that is immediately obvious to a human. Automated reasoning with OWL can also draw conclusions that logically follow through a chain of reasoning, even when the conclusions are not obvious.

One kind of conclusion is determining that there is a logical inconsistency. This tells you there is a bug in your ontology or in your data. For example, in Figure 1.7 the red line with an X in it denotes that nothing can be both a person and a corporation. If we already know that Google is a corporation, and someone comes along and mistakenly says that Google is also an instance of person, there is a problem that an automated reasoner can detect. The computer has been explicitly told that:

1. nothing can be both a corporation and a person and

2. Google is both a corporation and a person.

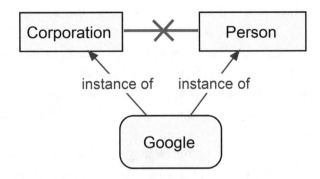

Figure 1.7: Reasoning helps to find an inconsistency.

This is an example of two kinds of things having no overlap. Look again at Table 1.1. See if you can find other examples of two kinds of things that cannot overlap.

Exercise 2: How do you resolve the apparent contradiction that the U.S. Supreme Court's Citizen's United decision declared that a corporation is legally a person with the above common sense example in the ontology? Can you think about it in such a way that there is in fact no contradiction?

1.8   DATA AND METADATA

To create a good model of the subject matter of interest, OWL needs to give modelers a way to say the following.

1. There are individual things.

2. There are kinds of things, some of which do not overlap.

3. An individual is an instance of a certain kind of thing.

4. There are more specific and more general kinds of things.

5. There are relationships between things.

6. Things have attributes with literal values.

These things must be said in a way that supports drawing conclusions both to add new information, and to detect and debug logical inconsistencies. They can be viewed as requirements for OWL, or any ontology modeling language, for that matter.

So far we have been very informal in describing different subjects. The diagrams I have been drawing are meant to be the kinds of diagrams you might draw on a whiteboard in a brainstorming session. Afterward, you would use them as the basis for creating an ontology using the more formal notation of OWL. That is covered in detail in the next chapter. We will now give a hint about what that will look like.

Once you have an ontology, it can be used to communicate meaning to other humans. While this is important in its own right, we will focus on how humans can get the computer to do useful things with the ontology. One of the main things you will want to do with an ontology is to use it as the basis for creating individuals and relationships between them and storing that as data. This is called populating the ontology.

The ontology provides a vocabulary for creating individuals and making statements about them. For example, consider the following statements in (not-too-stilted) English.

- John Doe *is an instance of* Person.

- Jill *is an instance of* Doctor.

- Jane *is an instance of* Nurse.

- John Doe's visit to the doctor *is an instance of* Patient Visit.

- John Doe's visit to the doctor was on date 12Jan2013.

- John Doe's visit to the doctor has care recipient, John Doe.

- Jill was a care provider on John Doe's visit to the doctor.

- Jane was a care provider on John Doe's visit to the doctor.

- John Doe received care from Jill.

- John Doe received care from Jane.

The vocabulary for the subject matter of healthcare is in gold for kinds of things, blue for relationships connecting individuals, and green for attributes with literal values. While not as natural-sounding, it is technically accurate to say that attributes are relationships that connect individuals to literals.

The individuals are in burgundy, and the single literal is in black. Note the generic relationship, *is an instance of*; it is neither a created individual nor part of the vocabulary of healthcare. Rather, it is part of the vocabulary for modeling. The formal name for this in OWL is `rdf:type`. We will get into that in the next chapter.

Each of the above sentences in English has a subject, a predicate (i.e., verb), and an object and is asserting something to be true. Count the parts: one, two, three—each sentence is a triple. Some example triples are graphically depicted in Figure 1.8.

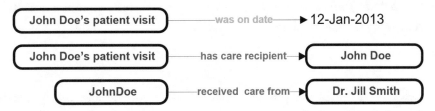

Figure 1.8: Assertions as triples.

So the ontology is the vocabulary for talking about the subject matter of interest. It is used to create and give meaning to data. That vocabulary is also represented as triples. For that reason the ontology is said to play the role of metadata for a database of triples.

In the next chapter, we describe how OWL meets the six requirements (at the beginning of this section) for describing subject matter in general, and how to create data using the subject matter vocabulary.

1.9   SUMMARY LEARNING

In this chapter, we learned what kinds of things we need to say when building an ontology.

What Is an Ontology?

An ontology is a model of some subject matter that you care about and OWL is a formal language for expressing the ontology. It communicates what kinds of things there are and how they are related to each other in a way that supports automated reasoning. An ontology can also be used informally to communicate shared meaning among humans.

What Do You Need to Say When You Build an Ontology?

The kinds of things we need to say to build an ontology are relatively few. They constitute an informal set of requirements for OWL. What you need to say is that there are:

1. individual things;

2. kinds of things (some of which do not overlap);

3. individuals of a certain kind;

4. more specific things and more general kinds of things;

5. relationships that connect things to other things; and

6. relationships that connect things to literals.

The things that are said are called assertions, and they are represented as triples. The ontology provides a vocabulary that plays the role of metadata.

Drawing Conclusions

If we are careful to say things very precisely, then the computer can draw conclusions for us. The act of drawing conclusions is referred to as performing inference. Special computer programs do this; they are called inference engines or reasoners. Two ways that this helps are (1) it makes the computer seem smarter and (2) the reasoner can detect and explain logical inconsistencies which leads to better ontologies. Stating that two kinds of thing are not overlapping is a big help in spotting inconsistencies.

CHAPTER 2

How Do We Say it in OWL?

2.1   INTRODUCTION

In Chapter 1, we described the key things that need to be said in order to build a model of some subject matter and to create data based on that model. In this chapter, we describe how to say those things in OWL. In doing so, we transition from mostly non-technical language to the more technical language of OWL.

This chapter presents the approximately 30% subset of OWL that you will use 90% of the time when you create ontologies. We mostly illustrate the concepts in the healthcare industry, taking up where we left off in Chapter 1. We also use examples from other industries and subjects.

2.2   SAYING THINGS

At the beginning of Section 1.8 we identified six kinds of assertions that we need to make to create and populate an ontology in OWL. Do you remember what they are? It would be a good idea to memorize them.

You also need to be able to say that a particular literal is of a certain kind (e.g., a string, an integer or a date). Because OWL provides the ability to say this directly in a way that is independent of the subject matter being modeled, it is not one of the six kinds of assertions.

2.2.1   AN ONTOLOGY IS A SET OF TRIPLES

In natural language, many sentences declare that something is so; these are assertions. The basic sentence structure in English has a subject, a predicate, and an object. Table 2.1 gives examples for each of the six kinds of assertions, split up into the three constituent parts.

	Subject	Predicate	Object
	Table 2.1: Example assertions in somewhat natural language		
1	Google	*is an instance of*	*An individual*
2	Corporation	*is an instance of*	*A kind of thing*
3	Google	*is an instance of*	Corporation
4	Corporation	*is a special kind of*	Legal Entity
5	Google	is a subsidiary of	Alphabet
6	Google	has official name	"Google Inc."

The legend for fonts is the same as used at the end of Section 1.8.

1. OWL constructs: black italics is for things that are part of the OWL modeling language.

2. Subject matter (i.e., metadata)

    a. gold is for kinds of things

    b. blue is for relationships connecting individuals to individuals

    c. green is for relationships connecting individuals to literals.

3. Data

    a. burgundy is for individuals

    b. black plain font is for literals

This same structure is used by OWL. An ontology is a set of assertions and each assertion is represented as a triple with a subject, a predicate, and an object.

Thus, an ontology is represented as a set of triples.

1. **Subject:** The subject of a triple is the thing about which something is said.

2. **Predicate:** The predicate of a triple indicates the sort of thing that is being said; it corresponds to a way that two things can be related, or a way that something is related to a literal. Echoing English, it often corresponds to a verb.

3. **Object:** The object of the triple is the individual or literal that is linked to the subject via the predicate. When the object is a literal, it is often thought of as a "value."

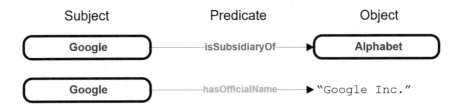

Figure 2.1: Assertions represented as triples.

This is extremely important and it bears repeating.

1. An ontology is a set of assertions.

2. Each assertion is represented as a triple with a subject, a predicate, and an object.

3. An ontology is represented as a set of triples.

Of course, OWL uses a much more precise and formal notation than the quasi-English sentences above. Table 2.2 has the formal triples that correspond to the assertions in Table 2.1.

A set of triples forms a directed graph—which is just a set of nodes with arrows connecting one node to another. The subjects and objects are the nodes and the predicates are the labels on the arrows. Each arrow corresponds to a single triple. The arrow goes in the direction *from* the subject *to* the object.

Notice that some of the entries in Table 2.2 appear in more than one triple (e.g., `doe:_Google` & `doe:Corporation`). The latter appears twice as the subject and once as the object. When represented as a graph, multiple occurrences snap together as one as in Figure 2.2. Graphs representing knowledge and data as depicted in Figure 2.2 are commonly referred to as knowledge graphs.

	Subject	Predicate	Object
	Table 2.2: Assertions formally encoded as OWL		
1	`doe:_Google`	`rdf:type`	`owl:NamedIndividual`
2	`doe:Corporation`	`rdf:type`	`owl:Class`
3	`doe:_Google`	`rdf:type`	`doe:Corporation`
4	`doe:Corporation`	`rdfs:subClassOf`	`doe:LegalEntity`
5	`doe:_Google`	`doe:isSubsidiaryOf`	`doe:_Alphabet`
6	`doe:_Google`	`doe:hasOfficialName`	`"Google Inc."^^xsd:string`

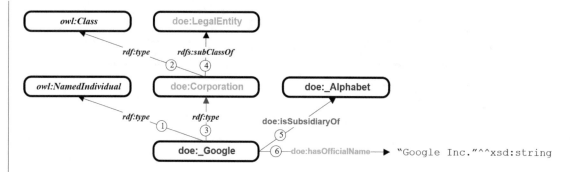

Figure 2.2: A set of triples forms a directed graph.

Each predicate is called a property in OWL. As depicted in Figure 2.1, there are two main kinds of properties:

1. *object property:* relates an individual to another individual and

2. *data property:* relates an individual to a literal

Literals should be thought of as being primitive—you cannot say anything about them. As such, OWL does not permit a literal to be the subject of a triple. This bears repeating: *a literal cannot be the subject of a triple.*

The two triples from Figure 2.1 are depicted as triples numbered 5 and 6 in Figure 2.2. Notice how even though Google is the subject of several different triples there is just one node for Google. That's because it is just one thing. There is no need to depict it more than once. Collapsing together different occurrences of the same node is sometimes called "node-folding."

We are using the following naming conventions for individuals, classes, and properties.

1. Classes: upper camel case (e.g., `LegalEntity`)

2. Properties: lower camel case (e.g., `isSubsidiaryOf`)

3. Individuals: leading underscore (e.g. `_Google`)

Note that other than the literal in the extreme lower right, all the entries in Table 2.2 have three parts. There is a short prefix, a colon, and the main term. We look into this further in the next section.

Note also that some of the triples help to define the subject matter (2 and 4) and others represent data relevant to that subject (1, 3, 5, and 6). When building an ontology, the focus is on the subject matter. The main work is to create and specify the meaning of classes and properties. In addition, it's not unusual for there to be a few special individuals that help to define the subject matter. For example, suppose Google was building an enterprise ontology and wanted to define what an internal organization is (from its perspective). An internal organization would include Google itself, or any organization within Google. To define this requires referring to an individual that represents Google itself. We give an example of this in Section 5.4.8.

Normally we use the term "ontology" to collectively refer to the triples representing the subject matter. In this sense, we can think of the ontology as being metadata; it serves the role of a data schema and gives meaning to the data. For example, triple 3 says Google is a corporation and triple 4 tells us something about what it means to be a corporation.

The fact that there are triples that represent both the metadata and the data is a very important feature of OWL that is unlike a traditional relational database. Because it's all triples, the data schema and the data go into the same data store—called a Triple Store.

Finally, notice that the literal for Google's official name indicates the kind of literal it is, in this case a string. The technical term in OWL for "a kind of literal" is `rdfs:Datatype`. Next we examine how terms such as `doe:_Google` and `rdfs:Datatype` are constructed. Table 2.3 summarizes the transition from informal English to formal OWL.

Table 2.3: Saying things in English and in OWL			
	Kind of Thing to Say	**Example of Saying It.**	**OWL Construct Used**
1	There are individual things	Google is an individual	(an) `owl:Thing` & (an) `owl:NamedIndividual`
2	There are kinds of things	Corporation is a kind of thing	(an) `owl:Class`
3	An individual is an instance of a certain kind of thing	Google is an instance of Corporation	`rdf:type` links an individual to a class
4	There are more specific and more general kinds of things	A Corporation is a specific kind of Legal Entity.	`rdfs:subClassOf` links two classes
5	There are relationships between things	Google is a subsidiary of Alphabet	an `owl:ObjectProperty` links two individuals
6	Things have attributes that relate them to literals	Google's official name is "Google Inc"	an `owl:DatatypeProperty` links an individual to a literal

2.2.2   NAMESPACES, RESOURCE IDENTIFIERS, AND OWL SYNTAX

There is some new notation in Table 2.2. Except for the one literal, every subject, predicate, and object is of the form `xxx:yyy`. Each of these is a globally unique identifier specifically designed for the Web: it is called a "uniform resource identifier" (URI). A "resource" is what the URI is identifying. For example: `doe:_Google` identifies the company, Google; `owl:Class` and `rdf:type` identify constructs in the OWL language. These are actually abbreviations using XML namespaces (see Figure 2.3). For example, "`rdf:type`" is the abbreviated form for the full URI which is:

`http://www.w3.org/1999/02/22-rdf-syntax-ns#type`.

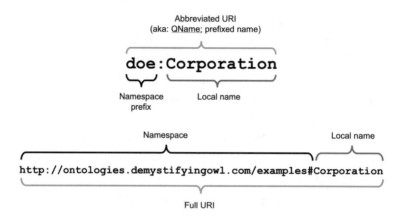

Figure 2.3: A namespace prefix is an abbreviation.

The term "qualified name" (QName[2] for short) is often used to refer to a URI abbreviated in this way. The term "prefixed name" is also used. The part before the colon is called the namespace prefix and the part after the colon is called the local name or sometimes a fragment identifier. A namespace gives the ability to have independently governed and managed ontologies without terminology clashes. For example, you might have an ontology for finance with the term `fin:Bank`. An ontology about river ecosystems might have the term `riv:Bank`. The different namespace prefixes distinguish the vocabulary of terms related to finance from the vocabulary describing river ecosystems. A namespace corresponds to a subset of URIs and can be thought of as vocabulary. Each term in the vocabulary starts with the same namespace prefix.

The OWL language makes use of multiple vocabularies. The main ones that you need to be familiar with are listed in Table 2.4. This is because OWL was not built from scratch, it reused existing vocabularies. OWL has its own namespace, but the OWL language uses constructs from three other namespaces: `rdfs`, `rdf`, and `xsd`. For now, think of RDFS as a subset of OWL and think of RDF as the language that OWL uses for representing triples.

OWL borrows specific datatypes (like string and integer) from XML Schema. The table lists one additional namespace, for SKOS, which stands for Simple Knowledge Organization Systems. It was designed to represent taxonomies and thesauri in OWL syntax. A few terms from this namespace are coming into common use in ontology development. We will say a bit more about this in subsequent chapters.

---

[2] https://en.wikipedia.org/wiki/QName

Table 2.4: Some common namespaces		
Vocabulary	**Prefix**	**Long Form**
OWL 2	owl	`http://www.w3.org/2002/07/owl#`
RDF Schema	rdfs	`http://www.w3.org/2000/01/rdf-schema#`
RDF	rdf	`http://www.w3.org/1999/02/22-rdf-syntax-ns#`
XML	xml	`http://www.w3.org/XML/1998/namespace`
XML Schema	xsd	`http://www.w3.org/2001/XMLSchema#`
SKOS	skos	`http://www.w3.org/2004/02/skos/core#`

When you create your own ontologies, you will create and use one or more of your own namespaces (with corresponding prefixes) for the vocabulary of terms that are in your ontology. In the above examples, we use the prefix "`doe`" which is meant to suggest: Demystifying OWL for the Enterprise. The expression "`doe:Corporation`" is short for

`http://ontologies.demystifyingowl.com/examples#Corporation`.

The namespace itself contains everything up to and including the "#".

Note that a namespace prefix is local to a particular file and although it is unusual to do so, two different files can specify two different prefixes for the same namespace. For example, there are two commonly used prefixes for the XML Schema namespace. The one listed above is `xsd`, but `xs` is also commonly used. The parser processing the file for use in a target system substitutes in the full namespace when constructing the URI, as depicted in Figure 2.3.

Even though some namespace prefixes are pervasive (e.g., `rdf`, `rdfs`, and `owl`) they still have to be present in each file. Typically, tools will add the standard namespaces for you. See the top several lines of the OWL in Figure 2.5. It would be technically possible to use other prefixes for these common namespaces or to use these standard namespace prefixes for other namespaces, but it would be perverse to do so.

URIs vs. IRIs

OWL uses IRIs,[3] a more general version of URIs for international use. It supports Unicode characters, which include any special characters or scripts in other languages. In case you ever want to use bowl of spaghetti as a namespace prefix, you can. There is a Unicode character for that.[4] For the remainder of this book, we will mostly use the more general term "IRI" rather than "URI."

---

[3]   Internationalized Resource Identifier: http://en.wikipedia.org/wiki/Internationalized_resource_identifier.
[4]   http://www.snee.com/bobdc.blog/2016/06/emoji-sparql.html.

OWL Constructs and Expressions

OWL is a language for representing ontologies. A natural language such as English is made up of words and punctuation. Words are combined to make noun phrases and verb phrases, which in the simplest case are individual nouns and verbs. Noun and verb phrases combine with other kinds of words to make sentences.

An OWL construct is analogous to a word. OWL constructs are combined to make OWL expressions. A triple is an assertion, and corresponds to a typical sentence in English.

OWL Syntax

There are a number of OWL syntaxes in common use. The main ones are listed below and will be introduced later in this chapter. Depending on the point we are trying to make we will use a variety of different syntaxes and notations.

- *RDF/XML* is the original OWL syntax, but is not very easy to read. It is still in widespread use, but is growing less and less popular. It is not used in this book.

- *Turtle* makes the triples nature of OWL more obvious. It is emerging as the de facto standard.

- *Manchester syntax* is the most concise and was designed for human readability.

- *e6Tools* is a compact visual syntax

2.2.3   SUMMARY: INFORMAL TO FORMAL

The following summarizes the shift from informal to formal terminology that takes place in moving from Chapter 1 to Chapter 2.

1. Rather than speaking of "building a model of some subject matter," we simply say "building an ontology."

2. Rather than speak of "things," "kinds of things," and "generalization/specializations" we speak of "`Individuals`," "`Classes`," and of "`subClassOf`" relationships, respectively.

3. Rather than saying that something is "some kind of thing," we say an `Individual` is an "instance or member of a Class," or, equivalently, that the "individual's type is some Class."

4. Rather than saying that a literal is of a certain kind, we say that literal "has a datatype."

5. Rather than speaking of "relationships between things," we will speak of "object properties."

6. Rather than speaking of "attributes of things," we will speak of "data properties."

7. Rather than speaking of "kinds of literals," we will speak of "datatypes."

8. Rather than speaking of "saying something," we will speak of "asserting a triple."

9. Instead of speaking of "drawing conclusions" to add new information, we will speak of "inference," which adds new triples.

2.2.4 NOTATIONAL CONVENTIONS

Throughout this book we will frequently be referring to generic concepts in English, independently of how they will be represented. These will generally appear in an ordinary font, even if they are later represented as classes or properties. OWL syntax will appear using a `fixed font`. In paragraphs of text when I refer to properties or classes I will generally leave out the namespace and leading colon. For example, I might say that a `HealthCareProvider` connects to a `PatientVisit` using the `careProvider` property.

When OWL syntax is set apart from regular paragraphs, such as in a set of triples, or in figures or tables, I either use a namespace prefix with a colon, or where brevity counts, just a leading colon. That serves as a reminder that the full IRI needs to have more than just the local name, even if it is not being depicted. For example:

```
:_JohnDoeDoctorVisit12Jan :careProvider :_DrJillSmith
```

2.3 A SIMPLE ONTOLOGY IN HEALTHCARE

2.3.1 HEALTHCARE DATA

Figure 1.2 contains an informal diagram that sketched out a few of the key components of what an ontology in healthcare would require. Figure 1.4 showed some sample data. These are just pictures created in a drawing tool that does not understand OWL. However, these diagrams depict preliminary requirements for a healthcare ontology. To create an ontology, you need to have an OWL authoring tool, often referred to as an ontology editor. In principle, all you need is a text editor and sufficient understanding of OWL syntax. In practice, most people use a tool with a graphical UI which saves out a file in valid OWL syntax.

Throughout this book we will use a variety of notations and syntaxes. Diagrams are easy to look at and understand. For that, we use a very compact and fairly self-explanatory notation from a proprietary Visio plugin called e6Tools designed by Simon Robe. It is used to recast the informal diagrams from Chapter 1 into a formal diagrammatic syntax that is designed to be an alternative OWL syntax for easier viewing. We will also show portions of ontologies in Turtle or Manchester syntax.

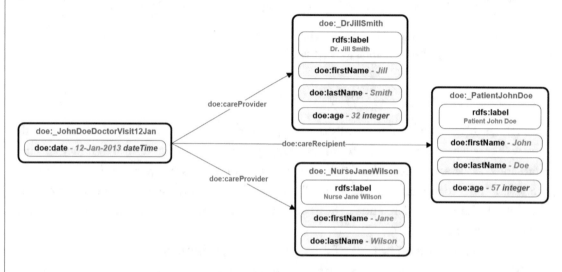

Figure 2.4: A patient visit in e6Tools OWL notation.

Figure 2.4 depicts a patient visit where both a doctor and a nurse were involved in providing care. We created it by using material from the diagrams in Figures 1.1, 1.2, and 1.4 and re-drawing it in Visio e6Tools syntax (henceforth abbreviated as "e6"). Notice the strong similarity between the boxes in Figure 1.2 depicting Jill Smith and John Doe and the e6 diagrams. E6 notation is explained below and summarized in Appendix A.2.

Exercise 1: First, see if you can determine which of the six kinds of assertions listed Table 2.3 are being made in Figure 2.4 and which are not. Pay particular attention to whether, where, and how object properties are used vs. data properties.

Recall that the former relate individuals to other individuals, and the latter relate individuals to literals. Here is a summary:

- Instances of: `owl:Thing`

    ○ `doe:_JohnDoeDoctorVisit12Jan`

- ◦ `doe:_DrJillSmith`

- ◦ `doe:_NurseJaneWilson`

- ◦ `doe:_PatientJohnDoe`

- • Instances of: `owl:ObjectProperty`

  - ◦ `doe:careProvider`

  - ◦ `doe:careRecipient`

- • Instances of: `owl:DatatypeProperty`

  - ◦ `doe:firstName`

  - ◦ `doe:lastName`

  - ◦ `doe:age`

  - ◦ `doe:date`

The diagram is very compact, but there is a lot going on. Each of the rounded boxes depicts an OWL Individual. At the top of each is the (globally unique) IRI for that individual. The convention we use for this book is for the IRI of an OWL Individual to have a leading underscore. There is an optional `rdfs:label` which is a human-readable name for the individual. It is an example of an annotation (described more fully in Section 4.10).

Each arrow corresponds to a triple that connects one individual to another. The label on the arrow is the name of the object property—more specifically, it is the IRI for that property. The subject is at the beginning of the arrow, the object is at the end of the arrow. Note how `careProvider` is used twice. These triples are called object property assertions.

There are also several data property assertions. To make the diagram more compact, these are not shown using arrows. Instead, the box has the IRI of the data property (in black bold font) and the literal value in a green font. We adopt the widely used convention for naming properties, which is to use lower camelcase. If the datatype is string, it is not depicted in the diagram, otherwise it often will be (e.g., `dateTime` and `integer`). The boxes depicting data property assertions are light green.

A representative sample of the key assertions in the above e6 diagram is listed below, first in English then in Turtle syntax. The first four triples relate only to healthcare, the last two are about computer systems using OWL. The label says how to present information to the user. The last triple has information used by OWL that is automatically created behind the scenes. Table 2.5 shows how these assertions are represented in OWL as triples. Figure 2.6 shows a more complete set of triples for the example in Figure 2.4.

1. John Doe's January 12 doctor's visit has care recipient: John Doe.

2. Dr. Jill Smith is of age 32.

3. John Doe's first name is "John."

4. Nurse Jane Wilson's last name is "Wilson."

5. Nurse Jane Wilson has the human-friendly label: "Nurse Jane Wilson."

6. `doe:_NurseJaneWilson` has the type, `owl:NamedIndividual` (which is a subclass of `owl:Thing`).

Table 2.5: Example triples in Turtle syntax		
Subject (Individual)	**Predicate (Property)**	**Object (Individual/Literal)**
`doe:_JohnDoeDoctorVisit12Jan`	`doe:careRecipient`	`doe:_PatientJohnDoe`
`doe:_DrJillSmith`	`doe:age`	`32`
`doe:_PatientJohnDoe`	`doe:firstName`	`"John"^^xsd:string`
`doe:_NurseJaneWilson`	`doe:lastName`	`"Wilson"^^xsd:string`
`doe:_NurseJaneWilson`	`rdfs:label`	`"Nurse Jane Wilson"^^xsd:string`
`doe:_NurseJaneWilson`	`rdf:type`	`owl:NamedIndividual`

Recall that (except for literals) the subject, predicate, and object must each have a globally unique identifier. OWL uses IRIs. They usually take the form of a uniform resource locator (URL), which ideally points to a location on the web. In the example above, the IRI for nurse Jane Wilson is: `doe:_NurseJaneWilson`. Because this is not very nice looking for humans, it is common to add a label that is used to refer to the individual when displayed by software, e.g., "Nurse Jane Wilson."

```
1    @prefix doe: <http://ontologies.demystifyingowl.com/examples#> .
2    @prefix owl: <http://www.w3.org/2002/07/owl#> .
3    @prefix rdf: <http://www.w3.org/1999/02/22-rdf-syntax-ns#> .
4    @prefix xml: <http://www.w3.org/XML/1998/namespace> .
5    @prefix xsd: <http://www.w3.org/2001/XMLSchema#> .
6    @prefix rdfs: <http://www.w3.org/2000/01/rdf-schema#> .
7
8    doe:healthABox rdf:type owl:Ontology .
9
10   #    Object Properties
11   doe:careProvider rdf:type owl:ObjectProperty .
12   doe:careRecipient rdf:type owl:ObjectProperty .
13
14   #    Data properties
15   doe:age rdf:type owl:DatatypeProperty .
16   doe:date rdf:type owl:DatatypeProperty .
17   doe:firstName rdf:type owl:DatatypeProperty .
18   doe:lastName rdf:type owl:DatatypeProperty .
19
20   #    Individuals
21   doe:_DrJillSmith rdf:type owl:NamedIndividual ,
22                            owl:Thing ;
23                 rdfs:label "Dr. Jill Smith" ;
24                 doe:age 32 ;
25                 doe:firstName "Jill"^^xsd:string ;
26                 doe:lastName "Smith"^^xsd:string .
27
28   doe:_JohnDoeDoctorVisit12Jan rdf:type owl:NamedIndividual ,
29                                    owl:Thing ;
30                        doe:date "12-Jan-2013"^^xsd:dateTime ;
31                        doe:careProvider doe:_DrJillSmith ,
32                                    doe:_NurseJaneWilson ;
33                        doe:careRecipient doe:_PatientJohnDoe .
34
35   doe:_NurseJaneWilson rdf:type owl:NamedIndividual ,
36                            owl:Thing ;
37                 rdfs:label "Nurse Jane Wilson" ;
38                 doe:firstName "Jane"^^xsd:string ;
39                 doe:lastName "Wilson"^^xsd:string .
40
41   doe:_PatientJohnDoe rdf:type owl:NamedIndividual ,
42                            owl:Thing ;
43                 rdfs:label "Patient John Doe" ;
44                 doe:age 57 ;
45                 doe:lastName "Doe"^^xsd:string ;
46                 doe:firstName "John"^^xsd:string .
```

Figure 2.5: Turtle syntax.

2.3.2   HEALTHCARE METADATA

The healthcare data we have in the previous section can be created in OWL and loaded into any OWL software. Although we are thinking about the kinds of things there are, e.g., patient, person, doctor, nurse, etc., they are not represented yet. We have referred to the specific properties (careRecipient, firstName), but we have not defined them in any way. This is an important thing to be aware of. You can create and load data as a set of triples without defining the metadata. This is convenient when you are getting started, but it is bad practice to never have any metadata.

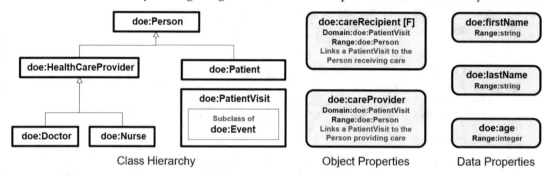

Figure 2.6: Healthcare metadata: classes and properties.

The metadata tells you about the data. In OWL, the metadata gives meaning to the data and supports inferences that can help find bugs as well as draw new conclusions. Figure 2.6 includes all the relevant metadata for the instances in Figure 2.4. In OWL, metadata mainly consists of:

1. classes for representing the kind of thing an individual is, e.g.,
   Person, PatientVisit

2. subClassOf relationship to indicate that a one class is a specialization of another class, e.g.,
   Patient is a subclass of Person, a PatientVisit is a subclass of Event

3. properties for characterizing the nature of the relationship between two individuals or between an individual and a literal, e.g.,
   careRecipient, firstName.

The figure shows classes as rectangles and properties with rounded corners. The data properties look pretty similar to the object properties. The former are green and the latter blue. There are three new things we have not yet seen. One is that we have added comments to explain the meaning in an informal way. The comments do not affect inference, but they are very important to help others understand your ontology. When ontologies get large, the comments will help remind you what you were thinking when you added that item to the ontology.

The second new thing is the [F] on the upper right of the `careRecipient` definition. This is a reference to the term "function" from mathematics. For our purposes it means that patient visit can only have one care recipient. This reflects the real world fact that in this particular healthcare organization, there is never more than one patient cared for during a single patient visit (at least not officially). Being functional is but one of several *characteristics* that we may specify for properties.

The other new thing is a way to say that properties are only relevant for certain types of subjects and objects. For example, whenever anyone enters an age, we want it to be an integer (ideally a positive one). In the ontology, we see string in the `firstName` and `lastName` properties; and we see integer in the `age` property.

Regarding the `careRecipient` property, we have in mind that the subject will be a `PatientVisit` and the object will be a `Person`. It would make no sense to say a `Regulation` had a `careRecipient`, or that the `careRecipient` was a `Disease`. So we want to specify two things:

1. What kind of thing must the Subject be, in a triple using a particular property? In OWL, this is called the "domain" of the property

2. What kind of thing must an Object be, in a triple using a particular property? In OWL, this is called the "range" of the property

If you do not find these terms very natural or intuitive, you have plenty of company. The terms come from mathematics and are used to describe functions, among other things. Think of it this way.

1. "`domain`" translates informally to "only applies to." If a property has domain C, then that means the property only applies to things of type C, e.g., `careRecipient` only applies to `PatientVisits`.

2. "`range`" translates informally to "range of possible values." If a property has range C, then the range of possible values that a property can have must all be instances of type C, e.g., the range of possible values for `careRecipient` is any kind of `Person`.

Being able to specify domains and ranges for properties is one of many ways to add semantic information to the ontology. The primary benefit is to remove ambiguity and make it easier for both people and machines to use the ontology. Domain and range can also be used to help the inference engine to spot errors.

It is important to be aware that OWL domain and range are not constraints. There is no gate keeper checking to make sure every care recipient is in fact a person. If someone accidentally says that the care recipient is an individual known to be an organization, then that organization is inferred to be a `Person`. Something similar happens for domain. Don't worry if this is a bit confusing. We go into more depth on domain and range in Section 4.4.

2.3.3   INDIVIDUALS AND THEIR TYPES

The astute observer will notice that there is one key thing missing. We have data representing several individuals, but did you notice what types they are? Look again at Figure 2.5. What are the types of the individuals? How helpful is that? The only types given are `owl:NamedIndividual` and `owl:Thing`. The ontologist does not specify these types, they are created by the system interpreting the OWL. Every individual is necessarily an instance of `owl:Thing`. Every individual that you give a name to (i.e., a IRI) is an instance of `owl:NamedIndividual`. We purposely created IRIs that are indicative of their meaning so we can readily create triples that explicitly indicate the types.

- `doe:_JohnDoeDoctorVisit12Jan`   `rdf:type`   `doe:PatientVisit`

- `doe:_PatientJohnDoe`   `rdf:type`   `doe:Patient`

- `doe:_NurseJaneWilson`   `rdf:type`   `doe:Nurse`

- `doe:_DrJillSmith`   `rdf:type`   `doe:Doctor`

This seems pretty obvious. We definitely want these individuals to be of the types indicated. However, it may not always be best to directly assign these types. Can you think of why that might be? Here is a hint. People can play many roles. A person might be a `Doctor` and a `Patient`, at different times. So one individual might belong to more than one class. Suppose a new class is defined in such a way that the person now fits that description too. For example, maybe there is a class: `FormerHospitalAdministrator` that Dr. Jill Smith belongs to because of a job she did in the past. It can be a lot of work keeping track of all the classes an individual belongs to. However, if we define our classes carefully, we can to some extent automatically classify individuals. In the next section, we will look at how we can perform this kind of inference.

Note that there are a number of different and more or less interchangeable ways to say in English what kind of a thing an individual is. The last one is closer to OWL, but less natural-sounding.

1. An individual is a instance of some Class.

2. An individual is a member of some Class.

3. An individual's type is some Class.

2.3.4   RICHER SEMANTICS AND AUTOMATIC CATEGORIZATION

We have said relatively little about the meaning of the classes. We have specified some subclass relationships and indicated how some of the classes are used with properties (via domain and range). But we can do much more.

Using Properties to Add Meaning to Classes

Consider a `PatientVisit`. We know that it is an `Event`. What else can we say about a `PatientVisit`? More specifically, what do we know to be true about every single `PatientVisit`, by definition? Think in terms of what relationships it must have with other individuals. If there is a `PatientVisit`, must there not be a `Patient`? Must there not also be a care provider of some sort?

Although OWL gives us a way to say this, it takes some practice to translate what you are thinking in natural language into OWL. One good way to do this, especially when initially learning OWL, is to first think of what you want to say and write it down in precise and natural English. After that, re-phrase it step by step into language that more directly fits the subject-predicate-object pattern using the classes and properties already defined. For example:

1. every `PatientVisit` has at least one `Patient`;

2. every `PatientVisit` is associated with one `Person` who is receiving care;

3. every `PatientVisit` has a `careRecipient` relationship with at least one `Person`; and

4. every `PatientVisit` is the subject of at least one triple where the predicate is `careRecipient` and the object is of type `Person`.

Exercise 2: Go through the same process starting with the statement "Every `PatientVisit` has at least one healthcare provider."

You should end up with something very similar.

- Every `PatientVisit` is the subject of at least one triple where the predicate is `careProvider` and the object is of type `Person`.

The last sentences in each case are directly expressible in OWL. Most OWL syntaxes for representing this kind of thing are hard to understand. This is one reason we draw pictures. One exception is Manchester syntax,[5] which is depicted on the right in Figure 2.7. On the left is the e6 visual notation. Some of it is surprisingly close to understandable (albeit stilted) English, the rest needs some explanation. What Figure 2.7 says about `PatientVisit` is:

1. `PatientVisit` is a Class

2. Every `PatientVisit` is an Event

3. Every `PatientVisit` necessarily has a care provider that is a Person

---

[5] See: https://www.w3.org/TR/owl2-manchester-syntax/.

4. Every `PatientVisit` necessarily has a care recipient that is a Person

The last three statements list conditions that must be true for every member of the `Person` class. We sometimes refer to them as *necessary conditions*. Note that the graphical notation echoes the Manchester Syntax very closely. We see something new here. An oval that represents a property is *inside* a rectangle that represents a Class. This reflects the fact that we are using properties to specify the meaning of the class. The (N) that you see is used in e6 to mean necessary.

```
Class: doe:PatientVisit

    SubClassOf:
        doe:Event,
        doe:careProvider some doe:Person,
        doe:careRecipient some doe:Person
```

Figure 2.7: Adding necessary conditions.

There is one thing here that will not be immediately obvious. Namely, why is the OWL construct `subClassOf` used to connect the class `PatientVisit` to the expression "`doe:careProvider some doe:Person`"? Only a class can be a subclass of another class. Therefore, the expression "`doe:careProvider some doe:Person`" must represent a class. But what class is that? Unfortunately, this one is pretty hard to work out, because it is not intuitive. If you are the adventurous type, see what you can come up with before reading on.

The class expression, "`doe:careProvider some doe:Person`" represents "the set of all things that have at least one care provider that is a person." To understand why the subclass construct is used we will again develop a series of sentences that are equivalent in meaning, like we did above. We will start with a simple subclass relationship.

1. [The class] `PatientVisit` is a subclass of `Event`.

2. Every [instance of] `PatientVisit` is an `Event`.

3. The set of all `PatientVisit`(s) is a subset of the set of all `Event`(s).

4. IF X is a `PatientVisit`, THEN X is also an `Event`.

5. Every [instance of] `PatientVisit` is *necessarily* an `Event`.

This is fine for ordinary named classes. However, "`doe:careProvider some doe:Person`" is a class *without* a name; it just is an expression. It can be very handy to not have to give things names, but it is very important to be *able* to think of a name that reflects the meaning.

Such names will often be long and unwieldy, erring on being unambiguous about the meaning. So, in English, the expression "`doe:careProvider some doe:Person`" means "The set of all things that have a care provider that is a person." A long but accurate name for this class might be `ThingWithACareProviderThatIsAPerson`. We can now do the same thing we did for `PatientVisit` and `Event`. To see how this works, we will just replace the class `Event` with the class `ThingWithACareProviderThatIsAPerson` in the above lines numbered 1–5. We add one sentence at the end that reads better as English.

1. [The class] `PatientVisit` is a subclass of `ThingWithACareProviderThatIsAPerson`.

2. Every [instance of] `PatientVisit` is a `ThingWithACareProviderThatIsAPerson` .

3. The set of all `PatientVisit`[s] is a subset of the set of all `Thing`[s]`WithACareProviderThatIsAPerson`.

4. If X is a `PatientVisit`, then X is also a `ThingWithACareProviderThatIsAPerson`.

5. Every [instance of] `PatientVisit` *necessarily* is a `ThingWithACareProviderThatIsAPerson`.

6. Every `PatientVisit` *necessarily* has a care provider that is a `Person`.

This illuminates why the subclass relationship is being used to connect the `PatientVisit` to the class expression "`doe:careProvider some doe:Person`." Convince yourself that all of these sentences mean the same thing. The last sentence is what we started with, so we have come full circle.

Let us revisit the original purpose. We were trying to capture the fact that every patient visit necessarily has a care provider that is a person. To do that, we create the class expression: "`doe:careProvider some doe:Person`" and make `PatientVisit` a subclass of it. Because of the way subclass works, this does the trick (see Figure 2.8). A good way to visualize this is using Venn diagrams.[6] In the diagram, shapes depict sets and a shape within a shape depicts a subset relationship. Points inside a shape indicate members of the set. I hope this is at least slightly clearer than mud; but, if not, take heart—we will be seeing many more examples.

Perhaps you are wondering whether a class expression like: "`doe:careProvider some doe:Person`" has a name. Indeed it does, such an expression is called an *OWL restriction*. Because there are many meanings of the word "restriction," we will refer to an OWL restriction as

---

[6] Per common usage, the term "Venn diagram" is being used loosely to include Euler diagrams. If you are curious, see: https://www.cs.kent.ac.uk/events/conf/2004/euler/eulerdiagrams.html.

a "property restriction." This makes a certain amount of sense, in that it is mainly used to narrow down (i.e., restrict) the possible meaning of a given class. The use of the property restriction "doe:careProvider some doe:Person" on the class PatientVisit restricts the set of possible individuals that can be members of the class, PatientVisit. For example, if it is known that something does not have a care provider, then it cannot be a PatientVisit.

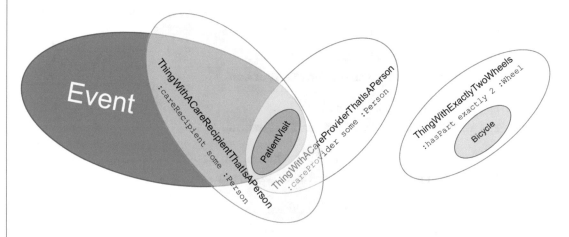

Figure 2.8: Property restrictions as sets.

In a similar way, we could restrict the meaning of a class Bicycle, to include just those things that have exactly two wheels. Making the class Bicycle a subclass of such a property restriction would mean that if you know something has more than two wheels or less than two wheels, then we can rule out the possibility of it being a member of the class Bicycle. See the Venn diagram in Figure 2.8.

This part is intuitive enough. What is not very intuitive, from a terminology perspective, is that strictly speaking, a property restriction always represents a class. At times I will use the term "property restriction class" to emphasize this fact. There are two main things that make an OWL property restriction class different than other OWL classes we have seen so far. First, it is a class that is defined by what properties its members have. Second, it need not have a name, it can be used purely as an expression.

Whether or not to give a property restriction class a name is a modeling decision. When learning how to use property restrictions, it is a good idea to at least think of a name for the property restriction classes, such as ThingWithACareProviderThatIsAPerson or ThingWithExactlyTwoWheels. For experienced OWL users, the main reason to give a property restriction class a name is if the class will be used in a variety of situations.

Data Properties

For the most part, data properties can be used in the same ways that object properties are. They can be arranged in hierarchies, they have domains and ranges, and they can be functional. For example, you can ensure that individuals can have at most one last name by making the property `doe:lastName` be functional. A data property can also be used in a property restriction. To ensure that every person has a last name you could create a property restriction class such as "`doe:lastName some xsd:string`" and make `Person` a subclass of that class. The differences all arise from the fact that they connect individuals to literals, not individuals.

Exercise 3: If you had to give that class "`doe:lastName some xsd:string`" a name, what might it be? What needs to be true to be a member of this class?

Automatic Categorization

We could go further still in defining `PatientVisit`. We might decide that for our purposes in our healthcare organization, anything at all for which the following is true, always has to be a `PatientVisit`.

1. It is an `Event`.

2. It has a `careProvider` that is a `Person`.

3. It has a `careRecipient` that is a `Person`.

After all, what else could possibly have these three things be true about it and *not* be a `PatientVisit`? Let's suppose that if there are any counter-examples, that they are obscure edge cases that we can ignore for our business. What is the impact of doing this? If we come across any individual for which the above three statements hold, then we know it must be a `PatientVisit`. Thus, inference has the ability to automatically determine that an individual belongs in certain classes. Two things are going on. First, we still infer that:

IF: an individual is known to

1. be a member of the `PatientVisit` class,

THEN: we can infer the following new information:
1. It is an `Event`
2. It has a `careRecipient` that is a `Person`
3. It has a `careProvider` that is a `Person`

We could do this before. What we have added is the ability to go the other way around.

IF: an individual is known to:

    1. be a member of the class `Event` AND

    2. have a `careRecipient` that is a `Person` AND

    3. have a `careProvider` that is a `Person`

THEN: we can conclude that the individual is a member of the `PatientVisit` class.

See Figure 2.9. It is the same as Figure 2.7 except that class equivalence is used instead of subclass.

```
Class: doe:PatientVisit

    EquivalentTo:
        doe:Event
        and (doe:careProvider some doe:Person)
        and (doe:careRecipient some doe:Person)
```

Figure 2.9: Adding equivalent conditions.

As previously noted, people can play various roles. Being a patient is such a role, as is being a healthcare provider. Specifically, any `Person` that is a `careRecipientOn` some `PatientVisit` is a `Patient`. We just have to define `Patient` using one of these newfangled OWL property restrictions (see Figure 2.10). Notice that until now we have only been talking about the relationship between a `PatientVisit` and the `Patient` from the perspective of the `PatientVisit`, which points to the `careRecipient`. However, to define patient in the way we just described, we have to talk about that relationship from the perspective of the `Patient`.

Let us consider this idea of perspective using a more familiar example of a relationship in everyday life: being a parent. If Peter says he is the "parent of" Patricia, what does Patricia say? Essentially, we are looking at the relationship going the other way, or its inverse. Patricia might say she "has parent," Peter. So a good name for the "parent of" relationship going the other way is "has parent." If we think of this as triples, we are just flipping the subject and the object.

We now revisit our `PatientVisit` example. The name of the `careRecipient` relationship going the other way is `careRecipientOn`. These names are chosen so that when we write them down as triples; we can read them in a way that sounds a bit like English. The `PatientVisit` has `careRecipient` `Person`, the `Person` is a `careRecipientOn` a `PatientVisit`.

In the two examples we have just seen, the pair of names for the property and its inverse are relatively easy to come up with, and are very suggestive of their inverse relationship. An equally accurate name for the inverse of `parentOf` would be `childOf`. This sounds good, but you do lose the direct clue in the names that one is the inverse of the others. It is not always easy to think of good names for inverse properties.

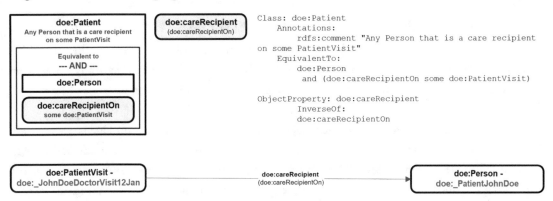

Figure 2.10: Equivalence using an inverse property.

In Figure 2.10 we show both the metadata and the data for this example. This is the same triple that we saw in Figure 2.4, the only difference is that we have added the name of the inverse property (non-bold in parentheses under the main property name). We have also removed some of the attributes to keep things simple. The data portion with the two linked individuals states that:

1. `:_JohnDoeDoctorVisit12Jan rdf:type :PatientVisit`

2. `:_PatientJohnDoe :careRecipientOn :_JohnDoeDoctorVisit12Jan`

3. `:_PatientJohnDoe rdf:type :Person`

Since we defined `Patient` to be equivalent to anything for which the above statements are true, the inference engine will automatically categorize John Doe to be of type `Patient`. We have a very similar situation to the patient visit example. First:

IF: we know an individual is a `Patient`,
THEN: we know that (1) it is a `Person` and (2) that is a `careRecipientOn` some `PatientVisit`.

Going the other way around:

IF: we know that an individual is (1) a `Person` and (2) a `careRecipientOn` some `PatientVisit`
THEN: we know the individual is a `Patient`.

That's very nice in principle. If you want to see what happens in practice, you can start using an ontology editor like Protégé[7] to create this simple ontology and single data triple and then run inference. This is illustrated in Figure 2.11. The top part of the figure shows a portion of a screen from Protégé with the inferences highlighted in yellow. In everyday English, the inferred assertions are:

1. John Doe is the care recipient on a particular patient visit (upper right of figure); and

2. John Doe is a patient. This is true because of what we said above (upper left).

The first inference is justified because the inverse triple is directly asserted. The second inference is justified by the second inference rule above. The tool itself will automatically generate an explanation that says what triples were used to conclude that John Doe is an instance of `Patient`. This is shown in the lower part of the Figure 2.11.

Note that an intermediate inference was made, in order to conclude John Doe is a `Patient`. Namely, if a relationship holds in the forward direction, the inverse relationship automatically holds. For example, if we say Peter `hasChild` Patricia, we can infer that Patricia `hasParent` Peter. The same thing is going on here. We assert the triple going in the direction from the `PatientVisit` to the `Person` receiving care, and the inverse relationship automatically holds. This is important because the definition of `Patient` used the inverse relationship: `careRecipientOn`.

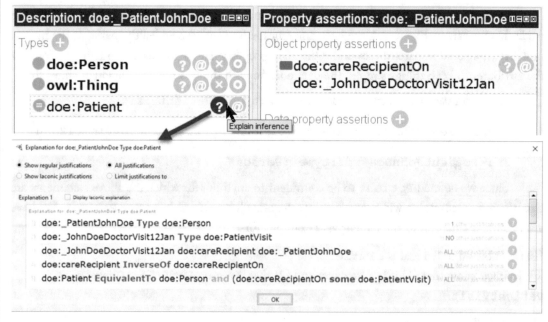

Figure 2.11: Inferring John Doe to be a `Patient`.

[7] http://protege.stanford.edu/ .

2.3.5   OTHER WAYS TO SPECIFY MEANING

There are a few more things that commonly arise in creating OWL ontologies. There are some additional ways to create classes from other classes. We can specify more meaning about properties and we can do some work to help the inference engine find simple mistakes.

Creating Classes from Other Classes

In the example above we saw how the class `Patient` was defined to be anything that was both a `Person` and that was a `careRecipientOn` some `PatientVisit`. Our graphical notation intentionally echoes this very closely. In a simplified Manchester syntax[8] it looks like:

```
Class: doe:Patient
EquivalentTo:
        doe:Person and
        (doe:careRecipientOn some doe:PatientVisit)
```

Let us consider some examples using a simpler variation of the same pattern. Suppose we needed to represent the class `PerformedProcedure`. Let's say we already have a class for `Procedure`, and another one for `HistoricalEvent`, where the latter is any event that has already happened. So a `PerformedProcedure` is something that is both a `Procedure` and a `HistoricalEvent`. In the world of vehicles, we might define the class `Motorcycle` to be something that is both a `TwoWheeledVehicle` and a `MotorizedVehicle`. The simplified Manchester syntax for `PerformedProcedure` is:

```
Class: doe:PerformedProcedure
      Annotations: rdfs:comment
                 "A procedure that has already been performed."
      EquivalentTo: doe:HistoricalEvent and doe:Procedure
```

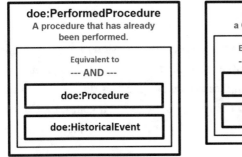
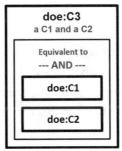
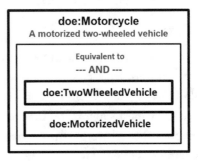

Figure 2.12: Class expressions using intersection.

The pattern is the same in all three cases. We are taking two base classes, (C1, C2) and combining them to define a third class, C3. Specifically, we are saying that a class C3 is equivalent to the expression "C1 and C2." To be a member of the third class, you have to be a member of C1 and a member of C2.

Our `Patient` example also fits this pattern. C1 is `Person`, C2 is the property restriction class "`careRecipientOn some PatientVisit`" that we might have called `ThingThatIsACareRecipientOnSomePatientVisit`, if we gave it a name. The set of all members of the class Patient, is precisely the set of all things that are members of both C1 and C2.

It turns out that the formal mathematical logic underpinning OWL regards every class as a set. The *and* operation we have just seen corresponds to set intersection. If we can have set intersection then we should of course also have set union. We do; it entails using an "*or*" operation in a completely analogous manner as we used "*and*" above. Below are two very common examples, one in healthcare and one for almost any enterprise. There is a third set operation that is used much less frequently corresponding to set complement. We will examine that in Section 5.3.2.

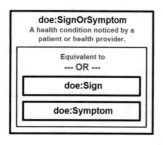 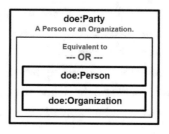

```
Class: doe:Sign
Class: doe:Symptom
Class: doe:SignOrSymptom
    EquivalentTo:
        doe:Sign
        or doe:Symptom

Class: doe:Organization
Class: doe:Person
Class: doe:Party
    EquivalentTo:
        doe:Organization
        or doe:Person
```

Figure 2.13: Class expression using union.

More Meaning for Properties

Subproperties

We can also add meaning to a property by saying it is a more specialized version of another property. In most large enterprises there are regular employees as well as a variety of contractors that do valuable work. There are important legal and tax implications regarding this distinction, so the ontology must support it. However, for some uses, you may just want to get a count, or a dollar amount that relates to persons that work for the enterprise, and you don't care whether they are contractors or employees, you want both.

To accomplish this, we start with a general property called `worksFor` that connects a person to the person or company that they do work for. We also need specialized properties that indicate whether the relationship is an employee/employer one, or whether it is about contracting. We might call these properties: `employedBy` and `contractorFor`. This means that:

IF: X :contractorFor Y   IF: X :employedBy Y
THEN: X :worksFor Y.   THEN: X :worksFor Y.

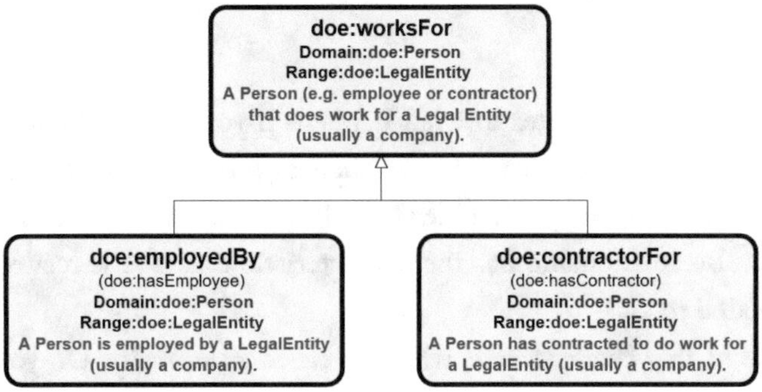

```
ObjectProperty: doe:worksFor
        Domain: doe:Person
        Range:  doe:LegalEntity

ObjectProperty: doe:employedBy
        InverseOf:    doe:hasEmployee
        SubPropertyOf: doe:worksFor
        Domain: doe:Person
        Range:  doe:LegalEntity

ObjectProperty: doe:contractorFor
        InverseOf:    doe:hasContractor
        SubPropertyOf: doe:worksFor
        Domain: doe:Person
        Range:  doe:LegalEntity
```

Figure 2.14: Subproperty hierarchy.

In general, for any property SP that is a specialized version of property P in the same way as in the `worksFor` example, we have:

IF: the triple <X SP Y> is asserted
THEN: the triple <X P Y> is necessarily true, and will be inferred.

Having a more specialized property is similar to having a more specialized class using the `subClassOf` relationship.

- IF: something is a member of the more *specialized class*
  THEN: it is also a member of the more *general class*.

- IF: the *specialized property relationship* holds between X and Y
  THEN: the *general property relationship* also holds between X and Y.

So, just as we say the specialized class is a `rdfs:subClassOf` the more general class, we also say the specialized property is a `rdfs:subPropertyOf` the more general property. This is why the same line and arrow head is used to mean both subclass and subproperty in the e6 notation.

Property characteristics

Recall the property, `careRecipient`, which points from a `PatientVisit` to a `Person`. We noted that for the healthcare company we are modeling, there is only ever one person that is receiving care on a `PatientVisit`. We saw that OWL gives us a way to say that in the ontology. Specifically, we say that the property is *functional*. Most properties are not functional, but many are. Every time you create a property, you should decide whether it is functional. This is but one of several characteristics that different properties may or may not have.

Exercise: 4: Can you think of other functional properties? Can someone have more than one first name? more than one last name? Can there be more than one person providing care on a given `PatientVisit`?

Another characteristic that commonly arises relates to what happens if you have a chain of triples linked by the same property. If you want to order people by their age, you might have an "older than" property. If Tommy is older than Jill, and Jill is older than Peter, then common sense tells us that Tommy is older than Peter. This relationship has the characteristic that if A is related to B and B is related to C, then A is related to C.

We have already seen an example of this characteristic, which is called *transitivity*. Do you remember it? We wanted the inference engine to have enough common sense to know that if a cancer patient is a subclass of patient, and if patient is a subclass of person, then a cancer patient is a subclass of person, without having to say it explicitly. In this case the transitive property is the OWL construct `rdfs:subClassOf`. Just like we can specify in OWL that a property is functional, we can also specify whether a property is transitive. In short, a transitive property is one that you can chain as many together as you want, and the property also holds from the beginning to the end of the chain. Consider some of the other properties we have seen, can you find any that are also transitive?

Just like you should ask whether every property is functional, you should also ask whether it is transitive. This is how the inference engine knows how to make the right inferences. You use whatever OWL constructs are available to express the semantics of the things you are creating. The ontology gives meaning to the data, and adding more meaning to the ontology adds more meaning to the data.

Exercise: 5: Is the property, `careRecipient`, transitive? Why or why not?

Common Techniques for Helping the Inference Engine Find Mistakes

Recall that there are two main benefits of inference. One is to generate new triples that logic and common sense tells you are true, without the need to write them down explicitly. This saves time up front; it also makes the ontology easier to maintain.

The other main benefit of inference is to help find mistakes. Some mistakes are very simple errors, not much more than a typo, others are much more subtle, and would be nearly impossible for a human to spot just by looking at the ontology. An example of a simple mistake would be to create an individual and put it in the wrong class. You might say an individual is a member of the class, `Patient`. Somewhere else, you or someone else might accidentally say that the same individual is member of the class `PatientVisit`. Common sense tells you that you cannot be both a `Patient`, which is a `Person` and also a `PatientVisit`, which is an `Event`. We want the inference engine to catch this mistake. We need a way to say that some combinations of classes do not have any members in common. This is another idea relating to sets. To say that there are no members in common means that the intersection is empty. In OWL we do this by saying that two classes are disjoint.

We want the inference engine to know that the two classes `PatientVisit` and `Patient` never overlap. There are far too many combinations of non-overlapping classes for it to be practical to manually make disjointness statements for each pair. It turns out we do not need to. We can go up to the more general classes like `Person` and `Event`, and just make those disjoint instead.

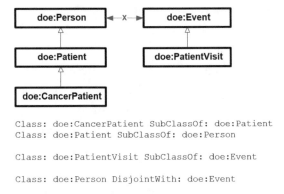

```
Class: doe:CancerPatient SubClassOf: doe:Patient
Class: doe:Patient SubClassOf: doe:Person

Class: doe:PatientVisit SubClassOf: doe:Event

Class: doe:Person DisjointWith: doe:Event
```

Figure 2.15: High-level disjoints.

Exercise: 6: Explain why this works. How can just specifying the two high-level classes being disjoint make the others also disjoint?

2.3.6   PAUSE AND REFLECT

Let's pause for a moment, because there is quite a lot going on here. For starters, you have just seen everything you will need for 90% of the time you spend building ontologies. The main effort now is to learn the ins and outs of using these constructs. There are quite a few additional features that you need from time to time, and we will introduce and explain them in subsequent chapters. Your initial focus should be to learn these core constructs cold. With experience, you will see the need for the more advanced features of OWL.

2.4   SUMMARY OF KEY OWL CONCEPTS AND ASSERTIONS

The core concepts in OWL are: individuals, classes, and properties. The ontology itself is a set of assertions (i.e., statements) of various sorts using the above core concepts. An assertion has a subject, predicate and object and is represented as a triple (see Table 2.2). Below we briefly summarize some key points about namespaces and then review the most common ontology assertions rendered in a simplified Manchester syntax of OWL, along with an English translation.

2.4.1   VOCABULARIES AND NAMESPACES

When you create an ontology you are creating a vocabulary of terms to define the subject matter of interest. It is much more than just a vocabulary, of course, because you are carefully defining the meaning of the terms using the logic-based OWL language.

You use the OWL vocabulary to create terms and expressions and give them meaning. Recall that OWL uses other vocabularies as well, mainly `rdf` and `rdfs`. After you create your own vocabulary of classes and properties, you can use it to create your data. Unlike most databases, the metadata and the data are expressed using the same representation (triples) and are loaded into and queryable from the same triple store.

You must create globally unique IRIs to identify the resources you create, chiefly classes, properties, and individuals. Most IRIs look just like the URLs we commonly use in web browsers. An IRI has a long name that is usually shortened for convenience using namespace prefixes. For example, `owl:Class` is the short form for the full IRI: `http://www.w3.org/2002/07/owl#Class` where "owl" is the namespace prefix, and "Class" is the local name of the IRI. They are joined by a ":".

Selecting names is important for understanding the ontology. They should be highly suggestive of their meaning, and any naming conventions that you choose should be used consistently.

2.4.2   INDIVIDUALS AND CLASSES

- **Individual:** An individual is a specific thing e.g., the house you live in, Joe Brown's health insurance policy, Janet Smith. The `rdf:type` construct is used to say what class an individual is a member of.

- **Class:** A class represents a kind of thing, e.g., building, health plan policy, patient. A class corresponds to the set of all things of that specific kind.

- **subClassOf:** A class is said to be a subclass of another class if it represents a more specific kind of thing than its parent class, e.g., a house is a specific kind of building. Every member of the subclass is also a member of the superclass. For example, if your particular house is a member of the class, `House`, then an inference engine can draw the conclusion that your house is also a member of the class, `Building`, without your having to say so.

The following are the main kinds of assertions relating to individuals and classes.

- Create a class:

  ○ `Class: Patient`
    Patient is a class referring to the set of all patients.

  ○ `Class: Person`
    Person is a class, referring to the set of all persons.

- Specify a subclass relationship:

  ○ `Class: Patient`
       `SubClassOf: Person`
    Patient is a subclass of Person, meaning that all patients are persons.

- Create an individual and give it a type:

  ○ `Individual: _JohnSmith`
       `Annotations:`
          `rdfs:label "John Smith"`
       `Types:`
          `Person`

  ○ The individual with unique identifier `_JohnSmith` is a member of the class `Person`, and uses the human readable label "John Smith."

- Set one class to be equivalent to another class

  ○ `Class: Person`
       `EquivalentTo:`
       `HumanBeing`

The two classes `Person` and `HumanBeing` have exactly the same members.

2.4.3   PROPERTIES

Property: A property is a way to relate individuals to each other or two literals. There are three main kinds as well as a way to specify subproperty relationships.

- **Object property:** An object property relates two Individuals,
  e.g., John `isParentOf` Peter.

- **Data property:** A data property associates a literal with an Individual,
  e.g., John `hasAge` 33.

- **Annotation property:** An annotation property is just like an object property or data property, except that they do not affect conclusions that the reasoner draws. They are used for such things as labels and comments, e.g., Building [has] comment "An enclosed structure that is to provide shelter."

- **subPropertyOf:** A property is said to be a subproperty of another property if it represents a more specific kind of relationship than its parent property, e.g., `employedBy` is more specific than `worksFor`.

Note that, strictly speaking, `rdfs:subClassOf` is also a property, but it is different from the rest in that it relates two classes rather than individuals or literals. This is OK because it is part of OWL itself. When we speak of properties we generally refer to properties created by the user, not those that are part of OWL. The following are the main kinds of assertions relating to properties.

- **Create an object property and specify its inverse:** for example: specify a way that two individuals can be related to each other from both perspectives,

    - ```
ObjectProperty: parentOf
ObjectProperty: hasParent
    InverseOf:
        parentOf
```

    - `parentOf` and `hasParent` are object properties. The first means that the subject is the parent of the object. The second is the inverse of the first, so it means that the object is the parent of the subject.

- **Create an object property assertion:** for example: assert that one individual has a certain relationship with another individual,

    - ```
Individual: _JohnSmith
    Facts: parentOf PeterJones
```

    - John Smith is the parent of Peter Jones.

- **Create a data property:** for example: specify a way that individuals can be related to a literal,

    - ```
DataProperty: hasAge
```

    - `hasAge` is a data property

- **Create a data property assertion:** for example: assert that an individual has an attribute expressed as some literal value.
```

- ○ `Individual: JohnSmith`
        `Facts: hasAge "33"`
- ○ The age of John Smith is 33.

- **Specify a subproperty relationship**

  - ○ `ObjectProperty: employedBy`
          `SubPropertyOf: worksFor`
  - ○ Being employed by someone is a more specific relationship than [just] working for them.

2.4.4   CLASS EXPRESSIONS AND RESTRICTIONS

OWL supports specifying meaning in a variety of ways by building up potentially complex expressions from the core primitives. The following are the main ways to create assertions using class expressions.

- **Define a class using set union**

  - ○ `Class: Party`
          `EquivalentTo:`
              `Person or Organization`
  - ○ Every member of the class `Party` is either a member of the class `Person` or it is a member of the class `Organization`. In addition, any member of either of the classes `Person` or `Organization` is also a member of the class `Party`. Set union here means logical "or."

- **Define a class using set intersection**

  - ○ `Class: Motorcycle`
          `EquivalentTo: TwoWheeledVehicle and`
                  `MotorizedVehicle`
  - ○ Every member of the class `Motorcycle` is a member of both of the classes `TwoWheeledVehicle` and `MotorizedVehicle`. In addition, anything that is a member of both of the classes `TwoWheeledVehicle` and `MotorizedVehicle` is member of the class `Motorcycle`. Set intersection here means logical "and."

- **Specify what properties members of a class must have.** By itself, creating a class and giving it a name doesn't tell us anything about the class. We can say more about a class by specifying what properties it must have.

○ `Class: Person`

    `SubClassOf:`

        `hasPart exactly 1 Heart`

○ Every member of the class `Person` has exactly 1 `Heart`. The expression "`hasPart exactly 1 Heart`" is a property restriction class that does not have a name. It means: anything that is in the `hasPart` relationship with exactly 1 member of the class `Heart`.

○ We can give a property restriction class a name using equivalence

  `Class: ThingWithExactlyOneHeart`

    `EquivalentTo:`

        `hasPart exactly 1 Heart`

The class `ThingWithExactlyOneHeart` has the same members as the set of all things that are in the `hasPart` relationship with exactly one member of the `Heart` class.

Notice that in the latter examples we made new classes by combining other classes and properties in useful ways. We created the class Party using the expression: "`Person or Organization.`" The expression: "`hasPart exactly 1 Heart`" is also a class; we can give it name to if we choose. There are other ways to create classes, e.g., using set intersection and complement.

What we did above may not be the most obvious way to say a `Person` has exactly one `Heart`—but that's the way it is done in OWL. Think of it this way. Saying that `Patient` is a subclass of `Person` says every `Patient` is a `Person`. Similarly, saying that every `Person` is a subclass of `ThingWithExactlyOneHeart` means that every `Person` is also a `ThingWithExactlyOneHeart`. Don't worry if you don't follow all of this now—it takes a while to wrap your head around it. Chapter 5 explains this in much more detail with many more examples.

2.4.5   DRAWING CONCLUSIONS

Reasoners, also known as inference engines, can draw new conclusions from the information that is asserted into the ontology. For example, if John Smith is a `Person`, and `Party` is defined as a `Person` or `Organization`, then the reasoner can conclude that John is a `Party`. If an individual is known to be a `Person`, the reasoner can conclude that it has exactly one `Heart`. Many other conclusions can be drawn.

One important use of drawing conclusions is to ensure that the ontology is logically consistent. For example, suppose we explicitly declare that nothing can be both a `Person` and a `Building`. If a mistake is made in creating the ontology, if an individual is found (by any combination of

direct assertion or inference) to be both a `Person` and a `Building`, the reasoner draws attention
to the problem and provides an explanation based on a trace of the reasoning that led to the error.

2.5   SUMMARY LEARNING

Saying Things with Triples

The first chapter identified what things need to be said when creating an ontology. This chapter
explained how to say those things in OWL. An OWL ontology is a set of assertions in the form
of triples, each composed of a subject, a predicate and an object. The subject is an individual that
has an IRI. The predicate is a property with an IRI. The object may either be another individual or
a literal. IRIs and namespaces are a key part of OWL. Anything with an IRI is called a resource.
There are various syntaxes for OWL, including Turtle and Manchester syntax.

We introduced formal terminology for the key concepts introduced in the previous chapter.

- `owl:Thing` is used to create individual things.

- `owl:Class` is used to create kinds of things

- `rdf:type` is used to say that an individual is an instance of a certain kind

- `rdfs:subClassOf` is used to say one kind is a specialization of another kind

- `owl:ObjectProperty` is used to create a way that two things can be related to
  another.

- `owl:DatatypeProperty` is used to create a way that something can be related to
  a literal.

Although everything is represented as triples, there is an important distinction between

- Data: the data you create by creating individuals and connecting them to other individuals and literals; and

- Metadata: the metadata which gives meaning to the data and supports inference.

Expressing Meaning

Meaning is expressed in various ways. There are class and property hierarchies that work very similarly. Properties may have domains which say what class the subject of a triple using the property
must be a member of. They have ranges which say what class the object of a triple using the property
must be a member of. Properties have certain characteristics, such as being functional or transitive.

There are numerous ways to create class expressions, chiefly including property restrictions and set operations like union (or) and intersection (and). Property restrictions are used to say what properties individuals of a given class have. One important use of inverse properties is for creating property restrictions.

Class equivalence can be used along with inference to automatically determine what classes an individual belongs to. That is an important difference between using subclass and class equivalence. The inference engine can also be used along with high-level disjoints to determine logical inconsistencies with an ontology.

Final Remarks

This completes our comprehensive introduction to OWL. Before we continue to delve more deeply into what we have learned, we will pause and consider some fundamentals on which OWL is based. That is the subject of the next chapter.

CHAPTER 3

Fundamentals: Meaning, Semantics, and Sets

In the first two chapters, we introduced the 30% of OWL that gets used 90% of the time. Everything was motivated by commercially relevant examples. You now have just enough knowledge to be dangerous. The goal of this chapter is to keep everyone safe. We step back and consider some fundamental concepts that are important for understanding the use of OWL for building ontologies. This will prepare you to dive into the details of OWL in subsequent chapters.

3.1 LOGIC

3.1.1 REASONING AND ARGUMENTS

OWL is a mathematically rigorous language that is based on formal logic. For our purposes, think of logic as the study of how to draw valid conclusions from existing information—or premises. A valid conclusion is one that logically and necessarily follows from the existing information. Two simple examples we have seen are:

IF: Every Corporation is a Legal Entity *and*
　　Google is a Corporation
THEN: Google is a Legal Entity.

IF: Every Cancer Patient is a Patient *and*
　　Every Patient is a Person
THEN: Every Cancer Patient is a Person.

The examples above instantiate general rules of inference and are very similar to what are called syllogisms, which go back to the time of Aristotle. The most famous syllogism is about men and mortality. First, we write it in the standard way that one normally sees a syllogism. Next, we explicitly separate out the premises from the conclusion, as above.

All men are mortal
Socrates is a man
Therefore Socrates is mortal

IF: All men are mortal *and*

Socrates is a man

THEN: Socrates is mortal.

Today we regard this as simple common sense, but the idea of having a formal structure for drawing valid conclusions was revolutionary in Aristotle's time. The philosophers who invented it first thought of it as having principled ways to construct valid arguments. This work was a precursor to modern logic.

Representation and Inference

There are various kinds of inference; the one we are concerned with is deductive inference (also called logical deduction). The above examples are instances of a pattern that is called an inference rule. The general pattern is:

IF: <set of premises>

THEN: <conclusion>.

> *In order for the computer to make inferences from existing information,*
> *there must first be a way to <u>represent</u> that information.*

There is a more compact way of saying IF <some premise> THEN <some conclusion> using the "=>" symbol, which means logical implication (not to be confused with equal to or greater than). For example, x => y is short for

IF: x

THEN: y.

What if the converse is also true? That is:

IF: y

THEN: x.

In this case, if either is true, then the other is also true, so they are basically equivalent. A common shorthand for this is to say that x <=> y. The symbol "<=>" means logical equivalence. Sometimes it is helpful to assert that something is *not* true. This is called negation. For example, suppose you knew that two classes are disjoint (like `Person` and `Event`). If you know that an individual is a member of the `Person` class, then you can definitively conclude that that individual is not a member of the `Event` class. In that case, you could say something like this:

IF: `:Person owl:disjointWith :Event.` and

`:_x rdf:type :Person.`

THEN: `¬(:_x rdf:type :Event).`

The symbol for logical negation (or "not" for short), is: "¬." It reads as "it is not the case that."

Each of the premises and the conclusion are assertions, which, in an OWL context, are represented as triples. Each triple represents information that is purported to be true, for the sake of a particular argument—that is to say, in the context of applying a particular inference rule. In triples, the inference rule relating to Google is expressed as:

IF: the following two triples are asserted:

```
:Corporation   rdfs:subClassOf   :LegalEntity.
:_Google          rdf:type       :Corporation.
```

THEN: infer the new triple:

```
:_Google          rdf:type            :LegalEntity.
```

This is an instantiation of the more generalized pattern that justifies an unlimited number of particular inferences like the one above:

IF: triples are asserted that match the following pattern:

```
:C1   rdfs:subClassOf   :C2.
:_x   rdf:type          :C1.
```

THEN: assert the new triple:

```
:_x   rdf:type          :C2.
```

Inferring the triple that _Google is of type LegalEntity is justified because the pattern matches with the following substitution:

- C1 = Corporation
- C2 = LegalEntity
- _x = _Google

Take a moment and see if you can depict what is going on here using a Venn diagram.

Axioms

Creating OWL ontologies consists of asserting a bunch of triples, each one representing what is believed to be true in the subject matter of the ontology. For example, the following two triples represent the facts that every patient is a person and that no person can also be an event, respectively:

```
:Person   rdfs:subClassOf   :Patient.
```

```
:Person   owl:disjointWith  :Event.
```

These asserted triples are often called "axioms." This term comes from mathematics. You may remember from high school geometry that an axiom is a starting point for drawing conclusions and proving theorems. An axiom does not need to be proved, it is assumed to be true.

Two Senses of the Term Logic

So far we have been talking about logic as the *study* of how to draw valid conclusions. In our context, a valid conclusion means it is OK to assert new triples based on certain other triples that are already asserted. However, the term "logic" has another closely related meaning: namely a particular *language* for representing information along with a set of inference rules that specify exactly what conclusions can be drawn from what premises. The first sense is logic *as a subject* (like biology or philosophy) the second sense is *a* logic *as a language* for representation and inference.

There are many different logics, each allowing different kinds of things to be said and each having a different set of inference rules. Even though each logic is different, many logics have a lot in common. A set of logics that share important characteristics may be grouped into a family of logics. OWL is an example of *a* logic, in the second sense of the word. There are several variants of OWL, the main one being OWL DL. OWL DL is a representation language belonging to the "description logic" family of logics. OWL variants will be discussed further in Section 7.7.

OWL is a subset of a more powerful logic called first-order predicate logic, often shortened to "predicate logic" or abbreviated as FoL for first-order logic. By "more powerful" we mean the ability to say a wider variety of things that support more inferences. Why would the W3C decide to use a weaker language such as OWL, rather than first-order logic?

The answer to this question is very important; it hinges on the results of ground-breaking research in the field of knowledge representation done in the 1980s. It was discovered that adding more expressive capability to a logic can seriously impact the inference performance. Some inferences will be impossible, and others will be hard to do efficiently at scale. A description logic has as a primary aim to have certain desirable inference properties that first-order logic does not have—this trades off the ability to say certain things.

Consider the following example. Let's say you have a class for whales and you want to use that class to say that the subject of a book is about whales. In triples, it might look like this:

doe:Whale	rdf:type	owl:Class
doe:Book	rdf:type	owl:Class
doe:_MobyDick	rdf:type	doe:Book
doe:_MobyDick	doe:isAbout	doe:Whale

It looks harmless enough, but a description logic such as OWL DL does not let you have a class as the object of a triple that uses a user-defined predicate. Why not? It is just one of several tradeoffs due to the fact that greater expressivity can cause inference capabilities to suffer. This and other limitations of OWL that you need to know about are described in Chapter 7.

3.1.2 FORMAL SEMANTICS AND SETS

A logic is said to have a formal semantics when there is a precise way to interpret each expression in the language. For example, how do we interpret the triples in the Moby Dick example above? Knowing how to do that is necessary in order to know exactly what inferences are sanctioned when. The formal semantics of first-order predicate logic (and OWL) is based on sets. This is why we use Venn diagrams to visualize some of the important inferences in OWL. Recall the inference asserting that Google is a Corporation.

IF: triples are asserted that match the following pattern:

```
:C1     rdfs:subClassOf    :C2.
:_x     rdf:type           :C1.
```

THEN: assert the new triple:

```
:_x     rdf:type           :C2.
```

Triples that represent true assertions but are left unstated are:

```
:C1     rdf:type           owl:Class.
:C2     rdf:type           owl:Class.
```

Formal semantics is about how to interpret OWL triples. This is done in terms of sets. You start by positing the existence of a bunch of individuals (logicians call this the "universe of discourse"). The individuals will be just those real or hypothetical things that you care about in the subject matter relating to your ontology. In healthcare the individuals will be people, patient visits, insurance claims, etc. Note that you can speak of all patient visits, or all corporations without being able to list them all. Statements in OWL will (implicitly or explicitly) refer to those individuals (e.g., _Google) or to sets of those individuals (e.g., Corporation).

OWL divides up the world into individuals and classes. Except in special circumstances, nothing can be both an individual and a class. Here are a few things we can say about some key OWL constructs in terms of sets.

- an owl:Class corresponds to a set. This correspondence is at the heart of what it means to have the formal semantics of OWL based on sets;

- rdf:type is used to say an individual is a member of the set corresponding to a particular class; and

- rdfs:subClassOf is used to specify that the set corresponding to one class is a subset of the set corresponding to another class.

For our particular example, this means that:

- C1 and C2 represent sets;

- the set corresponding to the class C1 is a subset of the set corresponding to the class C2;

- the individual corresponding to _x is a member of the set corresponding to class C1; and

- The individual corresponding to _x is a member of the set corresponding to class C2

The last statement is inferred from the meaning of subclass and subset. This is illustrated by the Venn diagram below.

Figure 3.1: The semantics of OWL is based on sets.

What is happening here is that we are interpreting things in the language of OWL (such as rdf:type and owl:Class, which are characters in a text file), in terms of sets. The formal semantics of OWL is the correspondence between constructs and expressions in OWL and what those constructs and expressions mean in terms of sets. For example, the expression Corporation corresponds to the set of all corporations, which in turn, is a subset of the set of all legal entities.

Each construct in OWL is best understood in terms of what inferences are justified when that construct is used. The rdfs:subClassOf construct (which means subset) in conjunction with the rdf:type construct (which means member of a set) is what justifies inferring the triple asserting that _Google is a LegalEntity.

The W3C published documentation that describes this in gory detail.[9] I occasionally consult it if I'm unsure about something. One day, you may too. Note that while it is technically correct to interpret the meaning of the triple

:C1   rdfs:subClassOf   :C2

as

"The set corresponding to the class C1 is a subset of the set corresponding to the class C2."

---

[9] https://www.w3.org/TR/owl-semantics/ *OWL Web Ontology Language Semantics and Abstract Syntax* https://www.w3.org/TR/owl2-syntax/ *OWL 2 Web Ontology Language Structural Specification and Functional-Style Syntax* (Second Edition).

we only need to be that verbose when discussing formal semantics. The correspondence between the OWL expressions and sets of individuals in a "universe of discourse" is precisely what formal semantics is about. Most of the time it will be OK to think of the class and the set as the same thing—i.e., think of C1 as being a subset of C2, as in Figure 3.1.

In OWL, everything boils down to sets of individuals and how those individuals are described and related to each other. In subsequent chapters, where we examine OWL's key constructs, we will explain what they mean in terms of sets.

Next, we look at the important idea of open-world inference.

3.1.3 THE OPEN WORLD

OWL uses open-world inference, an advanced topic. Many reasoners assume that the world is closed. If you ask a closed world reasoner to prove something and it cannot, it just says no. It assumes there is nothing else to know.

An open-world reasoner makes a different assumption—that there may be things that are true that it has not been told yet. If it received additional information, it might be able to conclude yes. Thus, if it is posed a question, it can say "yes," "no," or "I don't know."

A pithy way to understand the difference between open- and closed-world reasoners is that only the former can distinguish between "no" and "don't know." This has a profound impact on how to get inference working. Fortunately, you don't have to worry too much about it in the early stages of learning OWL. We will highlight those places where it is most useful to be aware of it and consider it again in Section 8.5.

Table3.1: Open- and closed-world reasoners			
	Yes	**No**	**Don't know**
Closed world	provably true	not provably true	n/a
Open world	provably true	provably false	neither provably true nor provably false.

3.1.4 RESOURCE IDENTIFIERS

An OWL ontology itself as well as its main components: classes, properties, and individuals are all called "resources."[10] Although it is a non-standard use of the English word "resource," that is what the "R" in RDF stands for. Each resource has an identifier called a universal resource identifier (URI). There is an international version called IRI which supports a wide variety of international languages. We will normally use the term IRI.

---

[10] https://www.w3.org/TR/1999/REC-rdf-syntax-19990222/#basic Accessed on May, 2018.

Just as your name identifies you, it is often convenient to think of an IRI as the name for the resource. The difference is while there may be several people with your name, there can only be one resource for a given IRI. This is important: an IRI uniquely identifies a single resource. Each IRI is composed of a namespace prefix and a local name (see Figure 2.3).

It is critically important to guarantee the uniqueness of the IRI. Failure to do so can corrupt your triple store. Using a namespace prefix that you control goes a long way towards guaranteeing uniqueness, but local names must also be unique. There needs to be an agreed individual or body that administers the names.

Unique Name Assumption

Although every IRI points to exactly one resource, the reverse is not true. Many systems assume that a single thing has a unique name. For OWL and the Semantic Web, the unique name assumption *does not hold*. A single RDF resource can be referred to by multiple IRIs. Consider the city of Edinburgh, Scotland. One of its IRIs is: http://dbpedia.org/resource/Edinburgh. Within the DBpedia namespace, it goes by other names as well, for different languages and also for nicknames, such as http://dbpedia.org/resource/Auld_Reekie, which means old smoke (from heavy burning of coal). Geonames also has an IRI for Edinburgh: http://sws.geonames.org/2650225/. All of these IRIs are also URLs that redirect to a human-readable webpage. Try it and see.

Figure 3.2: Auld Reekie (aka Edinburgh).

IRIs Should Persist

Another thing to consider with IRIs is that they should persist, ideally forever. A web address is an implicit contract, if people break that contract, then they are breaking the Web. So be sure not to include things in IRIs that you expect may change (e.g., organization or department names).

IRIs Should be URLs

Notice that the expanded form of an IRI looks just like a URL—that is by design. OWL was made for the Web. For example, the Good Relations ontology[11] has a class to represent business entities. It uses the namespace prefix "gr." The prefixed name, `gr:BusinessEntity` expands to http://www.heppnetz.de/ontologies/goodrelations/v1#BusinessEntity. Clicking on this link takes you to a web site that describes the meaning of the term.

Not everyone does it, but it is good practice for each IRI to also be a URL, so people and computers can go there and make use of the information about that resource. As another example, the IRI http://dbpedia.org/ontology/Bank takes you to a website which is depicted in Figure 3.3. This is what the human sees. There is also a mechanism called content negotiation[12] that allows the computer to access the same URL and get formal OWL definitions.

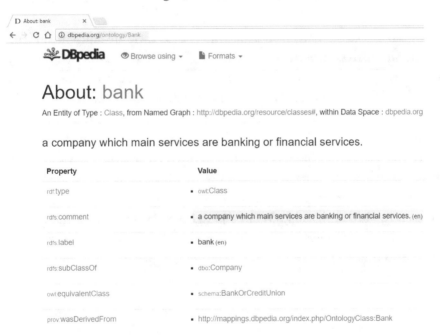

Figure 3.3: IRIs should be URLs.

Key points on resource identifiers are:

1. an individual IRI must be globally unique, it refers to exactly one resource;

2. many IRIs can refer to the same resource (no unique name assumption);

[11]   http://www.heppnetz.de/ontologies/goodrelations/v1 Accessed 11-05-2016.
[12]   https://en.wikipedia.org/wiki/Content_negotiation.

3. all names are IRIs. Good practice is to be a URL too;

4. namespaces correspond to subsets of IRIs managed by some governance body; and

5. ideally, never change an IRI, once it is published, it is like breaking a contract.

Next we take a closer look at literals and datatypes in terms of sets, comparing them to the sets view of individuals and classes.

3.1.5   LITERALS AND DATATYPES

In Chapter 2, we saw several examples of literals: "Wilson" and 32. Literals have types (such as string or integer) but other than that you cannot say anything about a literal; it's just a plain value. In technical terms, a literal cannot be the subject of a triple as it's always a dead-end node in a graph of triples.

Individuals and literals have many similarities. Just as there are different kinds of individuals that we divide into classes, there are different kinds of literals that are divided into datatypes (e.g., string, integer, dateTime). The datatypes are borrowed from XML. Pause for a moment to take a careful look at Table 3.2. It shows the direct comparisons and highlights the close analogy between individuals and classes on the one hand, and literals and datatypes on the other.

Before looking at Figure 3.4, test your understanding by representing the information in the left column of the table into triples. What happens if you try to make triples for the items in the right column? Then look at the figure to check your work. This is depicted in Figures 3.4 and 3.5, which illustrate these ideas from a set perspective. The ovals correspond to sets with their IRIs in boldface (e.g., for owl:Class and xsd:integer). An oval within an oval means subset. An item inside an oval means set membership (except for the boldface IRIs). The ones that are underlined are shown as ovals below.

Individuals and Classes	Literals and Datatypes
`owl:Thing` represents the set of all individuals	`rdfs:Literal` represents the set of all literals
`owl:Class` represents the set of all distinct *kinds* of individual	`rdfs:Datatype` represents the set of all distinct *kinds* of literal
`doe:Corporation` is an `owl:Class` It is a subclass of `owl:Thing` The set of all corporations a subset of the set of all `owl:Thing`(s)	`xsd:string` is an `rdfs:Datatype` It is a subclass of `rdfs:Literal` The set of all strings is a subset of the set of all literals
`doe:Corporation` represents a particular kind of individual, and corresponds to the set of all corporations	`xsd:string` represents a particular kind of literal, and corresponds to the set of all strings
`doe:Google` is a particular corporation, it is a member of the set of all corporations	"Google Inc." is a particular string, it is a member of the set of all strings

Table 3.2: Individuals and classes vs. literals and datatypes

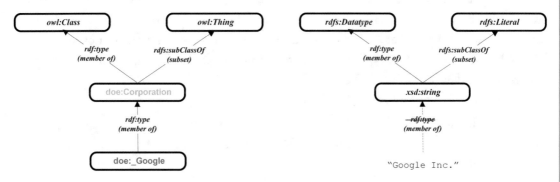

Figure 3.4: Classes and datatypes in terms of sets.

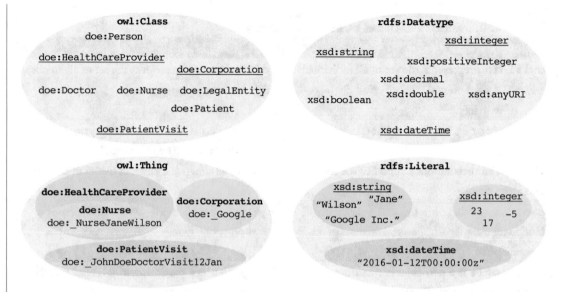

Figure 3.5: Literals and datatypes from a set perspective.

From a set perspective, the analogy between classes and datatypes is complete; however, there is a significant difference in terms of triples due to the stark divide in OWL between individuals and literals. The dotted arrow in Figure 3.4 connecting the string "Google Inc." to its datatype, `xsd:string`, indicates that the individual string, "Google Inc.", is a member of the set of all strings. But, there is never an explicit triple to represent the set membership relationship between a literal and its datatype; in particular, the following is not allowed:

`"Google Inc." rdf:type xsd:string .`

Why? Because a literal does not have an IRI and thus cannot be the subject of a triple. As we saw in Chapter 2, OWL uses another mechanism, e.g., `"Google Inc."^^xsd:string` is the Turtle syntax.

Commonly Used Datatypes

OWL provides a wide variety of datatypes. Most are borrowed directly from XML and their IRIs are in the `xsd` namespace. There are number datatypes including `integer`, `positiveInteger`, `decimal`, and `rational`. There are a few different string datatypes, but you mostly only care about `string` and the fact that there are language tags for multi-lingual applications. Language tags are discussed further in Section 4.10. There is a Boolean datatype for when you want to say something is either true or false. The datatype, `anyURI`, is for URIs and IRIs. There is a date-

`Time` datatype for representing a specific date and time with an optional time zone offset. Lastly, there are datatypes for representing binary data and XML literals.

The most commonly used datatypes are: `string`, `integer`, `dateTime`, `decimal`, `any-URI`, and `boolean`. There are a few dozen in total, but most are rarely used. [13]

3.1.6   METACLASSES

The observant and curious reader may have wondered why it is that if Venn diagrams can depict both set membership and subset relationships, then how come Figure 3.5 does not combine the two left ovals into a single diagram? How would that work?

Remember that in the Venn diagrams ovals are sets and non-ovals are members/instances of their containing oval(s). To represent the subset relationship between `doe:Corporation` and `owl:Thing` requires `doe:Corporation` to be an oval. To represent the set membership relationship between `doe:Corporation` and `owl:Class` requires `doe:Corporation` to be a non-oval. It is clearly impossible to represent it as both an oval and a non-oval at the same time. This is just a diagrammatic limitation that does not affect how you represent the same information as triples.

This situation arises when you have a class whose instances are themselves classes. That is what it means to be a metaclass. In this case the metaclass is `owl:Class`, which is part of the OWL language.

There is something else going on here that is a bit subtle. It is mistaken to think that `doe:Corporation` is a *subclass* of `owl:Class`. Instead, the following statements are true:

- `doe:Corporation` is an *instance* of `owl:Class` ; and

- `doe:Corporation` is a subclass of `owl:Thing`.

This is depicted in the upper-left and lower-left portions of Figure 3.5, respectively. In Chapter 7, we consider when it might be useful to create your own class whose instances are also classes. Can you think of any situations like that?

3.1.7   EXPRESSIONS

An important aspect of any expressive language like OWL is the ability to build up more complex expressions from basic constructs. Expressions in OWL are closely related to everyday expressions in arithmetic.

---

[13]   See: https://www.w3.org/TR/owl2-quick-reference/#Built-in_Datatypes for the complete list of OWL 2 datatypes. See: https://www.w3.org/TR/owl2-syntax/#Datatype_Maps for detailed documentation on all datatypes.

- Arithmetic expressions are built up from numbers and arithmetic operators such as addition, subtraction, multiplication, and division. The expression produces a new number. For example, "14 + 2" produces the number 16.

- Class expressions are built up from classes and operators on classes such as "and," "or," and "some." The expression produces a new class. For example, the expression ":Person or :Organization" results in a class that corresponds to the union of the sets corresponding to :Person and :Organization, respectively.

A single named class is a simple class expression. We saw the following examples of more complex class expressions in Manchester syntax in Chapter 2. Each expression refers to a class that may or may not have a name. For each, the meaning of the overall expression is derived from the meaning of the parts of the expression.

1. ":Person or :Organization"

    For an individual _x to be a member of the class denoted by this expression means that either _x is a member of Person *or* _x is a member of Organization. In terms of triples, it means that at least one of the following represents a true assertion:

    a. :_x   rdf:type   :Person.

    b. :_x   rdf:type   :Organization.

2. "doe:HistoricalEvent and doe:Procedure"

    For an individual _x to be a member of the class denoted by this expression means that _x is a member of HistoricalEvent and _x is a member of Procedure. In terms of triples, it means that both of the following represent true assertions:

    a. :_x   rdf:type   doe:HistoricalEvent.

    b. :_x   rdf:type   doe:Procedure.

3. "doe:careProvider some doe:Person"

    For an individual _x to be a member of the class denoted by this expression means that _x has a careProvider that is a Person. In terms of triples it means there is a triple whereby:

    a. _x is the subject of the triple;

    b. doe:careProvider is the predicate of the triple; and

    c. The object of the triple is a member of the class doe:Person.

The semantics of OWL is founded on sets. Here, the "and" and the "or" used in the above expressions correspond to set intersection and set union. These are called Boolean operators. The operator "some" is used to define property restriction classes.

Numbers and operators can be combined in many ways to form arbitrarily complex expressions such as "(3 * (14+2)/8)," which produces the number 6. Similarly, classes and certain OWL constructs that operate on classes (e.g., some, or, and and) can be used to form arbitrarily complex class expressions. For example, the expression, ((TwoWheeledVehicle and MotorizedVehicle) or MotorizedSkateBoard) produces the class that is the union of MotorCycle and MotorizedSkateBoard.

Compared to classes, there are very few expressions for properties—the main one that you may encounter is used to specify the inverse of an object property. These will be discussed in Section 4.5.1 as an alternative to introducing an explicit named inverse property like we saw in Figure 2.10.

Summary: Classes and properties can be simple, or they can be built up from pieces in expressions using various OWL constructs. There are many ways to create class expressions, and very few ways to create property expressions.

3.1.8   MEANING, SEMANTICS, AND AMBIGUITY

The word "semantics" means "meaning." We discussed the formal semantics of OWL by relating expressions in OWL to what they *meant* in terms of sets. The formal semantics tells the human how to develop an inference engine that, in effect, "knows" the meaning of OWL constructs such as rdf:type and rdfs:subClassOf, at least in the sense of producing all and only correct inferences.

If we want to give class expressions names, or more accurately, IRIs, we use class equivalence as follows:

1. doe:Party EquivalentTo

   (doe:Person or doe:Organization)

2. doe:PerformedProcedure EquivalentTo

   (doe:HistoricalEvent and doe:Procedure)

3. doe:ThingWithACareProviderThatIsAPerson EquivalentTo

   (doe:careProvider some doe:Person)

Now it can reasonably be said that the computer "knows" at least something about what it means to be a Party or a PerformedProcedure. Those meanings are in terms of the meanings of the OWL class operators and anything it knows about the components of the expressions. In this case the components of the expressions include:

- **Classes:** `Person`, `Organization`, `Procedure`, `HistoricalEvent` and

- **Object Properties:** `careProvider`

But so far, the computer does not know anything about these things. There are many things that could be said, but eventually things bottom out and what you have left are called primitives. In terms of formal semantics, all that can be said about the primitives comes from knowing whether it is a class, a property, or an individual. Below is one of the expressions color-coded to indicate what the computer knows about (green) vs. not (red).

`doe:Party EquivalentTo (doe:Person or doe:Organization)`

If you want to get a sense of how the computer sees this, I need only convert the red text to an unintelligible font.[14] It looks a lot different to you and me, but to the computer, both renderings are identical.

`doe:`ə̀˜‴ • ⌒ `EquivalentTo (doe:`ə̀◡‴ ⋯⁄↘ `or doe:`�‴ ◦˜\⤬◡˜•⤬⁄↘`)`

In fact, I don't even have to have the same characters in a different font, I can have different things entirely, and it is still the same to the computer.

`doe:SignOrSymptom EquivalentTo (doe:Sign or doe:Symptom)`

Let me emphasize. The computer cannot distinguish anything different among the prior three assertions on their own. I say "on their own" because if the computer saw **Person** several different places in the same ontology it would know that it referred to the same class on each occasion. We humans can be fairly certain that **Sign** and **Person** refer to different things, but the computer allows them to be the same. Why? Because OWL does not make the unique name assumption.

The important point is this. We humans read a lot into the names. It is easy to unwittingly assume that the computer knows what we know. Do not fall into that trap. An even more common trap is to assume that *other* people know what you mean by the terms you use. Take "person" for example. The U.S. Supreme Court says a corporation is a person. A biologist says a person has flesh and blood, which no corporation has. Thus, the term "person" is used to talk about two quite different concepts.

There is a tremendous amount of ambiguity in how people use words. Therefore, it is important to give good natural language definitions for all of the concepts in the ontology, and especially the primitive ones. While the inference engine won't be any wiser, it is an important signal to the humans that will be using your ontology. If they use it incorrectly, it will likely generate spurious inferences, or it will not generate the inferences you intended.

So, we have a second use of the term "semantics" that is very different from the meaning of the term "formal semantics." The latter term is very precisely defined by logicians and it is what

---

[14]   In this case "Bookshelf symbol 7."

relates OWL to sets. In the above example, we are using the term "semantics" in a less rigorous manner; it corresponds to the intended real-world meaning behind the concept you put in your ontology. Each triple that is asserted into the ontology about a given concept tells the computer more, and it narrows down the number of mistakes that can be made. This is a good thing, but only up to the point where that information is likely to be useful.

Summary: Ironically, the term "semantics" is itself ambiguous. We discuss two senses. The term "formal semantics" relates expressions in OWL to what they mean in terms of sets, but is agnostic to what objects and relationships in the real world you intend them to mean. That is the second sense of the term: "semantics." Sometimes it is called "real-world semantics." An inference engine is a powerful tool, but the old adage holds: "garbage in, garbage out."

We gave some examples of the different ways the term "meaning" is commonly used. The computer does not really know what things mean, it only knows what it is allowed to infer based on what it has been told. Natural language text is an important way to convey meaning to humans to ensure the ontology is used as intended. And although these text definitions will never participate in OWL inference, as natural language processing gets more and more powerful, having these definitions will become more and more useful for computing.

3.2   PRACTICAL MATTERS

3.2.1   CONCEPTS VS. TERMS

It is important to distinguish between (1) the underlying concepts that describe a subject matter and (2) the terms used to name these concepts. Focusing on terms is a common mistake. It is important to first get clear on the meaning of the concept you want to represent in OWL and then think of a good term for that concept.

The expression: "a rose by any other name is still a rose" translates in OWL to, "it makes no difference to the computer what the term is, what matters is the underlying concept that the term is naming." Communication between people is undermined by ambiguity, so choose terms that are hard to misunderstand. Even a term like "rose" is ambiguous. What is it that a woman might receive a dozen of on Valentine's Day? What is it that you buy from the nursery and go plant in your garden that one day might produce that Valentine's day gift? Ambiguity is everywhere, watch out for it.

- Rose: is it the plant or the flower or the species?

- Loan: is it the money or the contract?

- Risk: is it the bad thing happening or the probability of it happening?

3.2.2   THE WORLD OF TRIPLES

There are different ways to divide up the world of triples. The simplest and clearest distinction is between triples where the predicate is an object property vs. a data property. These are called object property assertions and data property assertions, respectively. In both cases the subject is an individual. For object property assertions, the object is an individual and for data property assertions the object is a literal. This is an important distinction to be aware of when modeling.

A second and entirely different way to divide up triples is based on whether they contribute to defining the subject matter being modeled, or whether they are populating the subject matter concepts with data. This closely corresponds to the difference between the metadata (or data schema) and data. The metadata defines terms such as `Corporation`, `Person`, and `isSubsidiaryOf`. Creating data consists of creating individual corporations and persons and relating them to other persons or corporations, among other things (see Figure 2.2).

Because the metadata defines the _Terms_ of the ontology vocabulary, it is sometimes referred to as the TBox. Because the data consists of _Assertions_ using the ontology vocabulary, it is sometimes referred to as the ABox. Unfortunately, the statements in the TBox are also assertions, so "ABox" is not the most helpful term. See Table 3.3 for a set of triples from the healthcare example that includes both metadata and data.

Table 3.3: Data/ABox and Metadata/TBox	
Metadata: Terms of the Ontology Vocabulary (TBox)	**Data: Assertions using the Ontology Vocabulary (ABox)**
`:Person rdf:type owl:Class`	`:_DrJillSmith rdf:type :Person`
`:PatientVisit rdf:type owl:Class`	`:_PatVisit32 rdf:type :PatientVisit`
`:careRecipient rdf:type owl:ObjectProperty`	`:_PatVisit32 :careRecipient :_JohnDoe`
`:age rdf:type owl:DatatypeProperty`	`:_DrJillSmith :age 32`

This second distinction is mostly syntactic. Are the triples creating and defining classes and properties or are they creating and relating instances to other instances or literals? However, there are some exceptions. For example, any company might wish to distinguish internal organizations from external ones. But if, say, Canon needed that class in their ontology, the definition of the class would need to refer to Canon itself, which is an individual corporation.

Another common exception arises for items in enumerated lists. For example, employees are categorized into exempt and non-exempt. There might be drop-downs menus needed for selecting among eye colors or age groups. In each case, although it is convenient to represent the individual items in the lists syntactically as individuals, they are really describing the subject matter.

Partly due to the existence of such special individuals, the terms TBox and ABox are not always used consistently. People will sometime ask: "should this individual go into the TBox?" when the more useful question is: "does this individual contribute to defining the subject matter." If the answer is yes, then it should go in the ontology where the classes and properties are being defined. In this book I will use the term TBox in the more inclusive way, which may include some special individuals. But be aware that others may use a more strict interpretation, excluding all individuals.

A third way that one can divide up triples is to distinguish those that are directly related to the real-world subject matter at hand, be it healthcare or finance vs. those that are really about a particular application related to that subject. We recommend keeping them separate because applications change much more rapidly than a real world subject. That, in turn, makes it easier to use and reuse the ontology over a longer period of time. Summarizing, we describe three different ways to divide up the world of triples:

1. datatype vs. object property assertions;

2. subject matter vs. data; and

3. real world vs. application.

While it is useful to be aware of the difference between data and object property assertions, they are freely mixed together in ontologies and datasets of triples. The second two distinctions call for separate subsets of triples. This leads us into the next section.

3.2.3   REUSE AND MODULARITY

Sooner or later when you are building ontologies, you will wonder whether someone else has already built an ontology that you can reuse. For example, the W3C has published ontologies on various subjects (e.g., organization, media, and provenance). Alternatively, you may wish to create ontologies that are specific to your business, but that will be used in a number of ways in your organization.

Also, when the ontology gets to be a certain size, you may wish to divide it up into separate modules, but also have a way to compose them all into the final ontology you need. Another situation when modularity is important is to separate out the data from the metadata. For example, a generic ontology for offers and pricing should be usable across many industries and products. It could be extended to include more specific concepts for a particular industry, or it could be used directly as a data schema for triples.

The good news is that OWL provides a simple mechanism that makes this reuse and modularity very easy to do. The construct is called `owl:imports`. When one ontology imports another ontology, the result is the same as if you had all the triples in each ontology in a single monolithic ontology. In other words, the resulting set of triples is the union of the sets of triples for each ontology. Although "`A imports B`" and "`B imports A`" results in the same set of triples, they are not the same, so import is directional.

- A `owl:imports` B means that when you use ontology A, you always get B's triples. You can still use ontology B on its own.

- B `owl:imports` A means you cannot use ontology B on its own. You always get A's triples. But you can use ontology A on its own.

The not-so-good news is that it is often difficult to find someone else's ontology that exactly suits your needs. Also, there are no hard and fast rules about how best to subdivide your ontology into modules. The only hard and fast rule about when to use import is, when you have an ontology that includes data, it should always import an ontology that contains the metadata. That way many different data sets can reuse the same metadata ontology (TBox). You don't have to reinvent all those classes and properties, and the inferencing can help find logical mistakes in your ontology. For example, look at Figure 2.15. If the classes `Person` and `Event` were in a separate ontology that you reused, then the disjointness axiom could spot the error if you accidentally said a single individual was both a `Patient` and a `PatientVisit`.

Figure 3.6 depicts an example of a network of importing ontologies. This could be for a company that provides materials to manufacturers of healthcare products. To run such a business, you need to be able to represent and store data about materials specific to healthcare products. There is a lot of chemistry that goes into materials, so you may want to import a separate ontology of chemistry concepts that is not specific to healthcare. To sell products, you need to have data about offers and pricing as well as the materials themselves. Materials must also have safety data which is managed separately from the price data, but it needs to know about materials, so it imports the materials ontology.

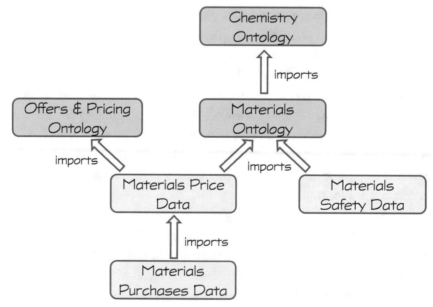

Figure 3.6: Network of importing ontologies.

3.2.4   TRIPLE STORES, QUERYING, AND SPARQL

The ontology is a set of triples and data is also represented as triples. We put them in a database that holds triples, called a triple store. Although we will not say a lot about it in this book, this is where the rubber hits the road. In a typical enterprise scenario, when an ontology is complete, you load it into a triple store and then populate it with data. A special query language called SPARQL was developed for querying a triple store. It is similar to SQL, and you can use it much the same way to get reports or to drive application functionality. Just like SQL, SPARQL returns a table as a result.

There are also important differences. First, the data is not stored in tables; it is stored in triples which collectively make up a graph. Second, the use of IRIs as globally unique identifiers eliminates the need to worry about joins or foreign keys. Another important difference is that both the data schema and the data are represented as triples. This means you can query the data schema as well as the data using the same query language. It also makes it possible to make changes to the data schema (the ontology) on the fly. This is not practical to do with a relational database. Next, we look at what we can do with querying both the metadata and the data in the same store.

You can query to find out what all the classes and properties are that make up the data schema (i.e., ontology). You can also run queries to ask which classes and properties are most often used. There is a growing number of triple stores that may be queried directly via the Web; these are called SPARQL endpoints. DBpedia is a very popular one; it is a database version of Wikipedia, where the ontology and the data are all in triples. If you go to the DBpedia SPARQL endpoint[15] and enter the following queries, you will find out which classes have the most instances and which object properties are used in the most triples:

```
SELECT ?c count(?i) WHERE {
  ?c rdf:type owl:Class.
  ?i rdf:type ?c.}
ORDER BY DESC(count(?i))
```

```
SELECT ?p count(?i) WHERE {
  ?p rdf:type owl:ObjectProperty.
  ?i ?p ?o.}
ORDER BY DESC(count(?i))
```

This casts illumination on what is important to those contributing to Wikipedia. Ignoring properties that link Wikipedia pages to each other and to external webpages, the most used property in Wikipedia is called "team." It connects an athlete to the sports team they are on. At the time of this writing, it is used nearly 2 million times (1,892,645, to be exact). Other popular properties

---

[15]   https://dbpedia.org/sparql, accessed January 15, 2018.

are `birthPlace`, `careerStation`, `starring`, `isPartOf`, and `country`. The classes with the most instances include: `Agent`, `Person`, `Place`, `Athlete`, and `Species`.

SPARQL Queries: A Closer Look

The SPARQL processing engine does two main things when executing a query. First, it identifies a subset of triples in the triple store that are of interest. The subset of triples is a subgraph of the whole graph in the triple store. Secondly, it returns just the information in those triples that is asked for, in the form of a table (just like SQL). The query itself must specify a pattern that declares what triples are of interest, and say what information should be returned. For example, say you wanted to know all the companies that are based in Chicago, IL and for each you wanted a table showing the official name of the company and the name of its CEO. The following query will give the correct result:

```
SELECT ?companyname ?ceoname
WHERE {
  ?company rdf:type doe:Company .
  ?company doe:isBasedIn doe:_Chicago .
  ?company doe:hasOfficialName ?companyname .
  ?company doe:hasCEO ?ceo .
  ?ceo      doe:hasName ?ceoname }
```

The subset of triples that are of interest is specified as a pattern in the WHERE clause. The pattern includes both constants and unbound variables. The unbound variables start with a question mark and are shown in bold. The query processor looks at all the triples in the triple store and finds those that match the pattern. One match that might be found is:

```
:_Boeing rdf:type doe:Company .
:_Boeing :isBasedIn :_Chicago_IL .
:_Boeing :hasOfficialName "The Boeing Company" .
:_Boeing :hasCEO :_Person_34521" .
:_Person_34521 :hasName "Dennis Muilenburg" .
```

One row in the table of data that is returned corresponds to one match of the pattern. The unbound variables after the SELECT in the query say which columns to have in the table. The row in the table of results that would correspond to this particular match is:

?companyname	?ceoname
"The Boeing Company"	"Dennis Muilenburg"

3.3   SUMMARY LEARNING

Logic, Semantics, and Sets

OWL is grounded in formal logic, which in turn is grounded in sets. Thus, sets are the foundation on which the meaning of every OWL construct is understood. Classes correspond to sets. Set membership is represented using `rdf:type` and `rdfs:subClassOf` means subset. A logic has a formal semantics that specifies what inferences are sanctioned under what circumstances based on what has already been asserted.

The meaning of each construct is understood in terms of the inferences that are sanctioned. The term "semantics" has two common meanings. One relates to the formalities of a representation language. The other is about what terms refer to in the real world. A common mistake is to read too much meaning into the names of the constructs. The computer only knows exactly what you tell it.

OWL expressions can be very simple, or more complex, being built up from simpler expressions and various OWL constructs.

OWL is an open-world system that can tell the difference between "no" and "don't know". Most systems are closed, in that if they cannot find something or prove that it is true they assume it is false.

Literals and Datatypes

When thought of as sets, there is an exact analogy that holds between:

1. classes and their individual instances and

2. datatype and literals of a given datatype.

However, the OWL language has a stark divide between individuals that have IRIs and literals that do not. As a consequence, `rdf:type` is not used to link a literal to its datatype. The most commonly used datatypes are for strings, numbers, and dates.

Concepts, Terms, and Resource Identifiers

Building an ontology is about identifying and representing the concepts and relationships in some subject matter of interest. You then create terms so you can refer to the concepts. The terms are important for communicating to humans, but they are utterly meaningless to the computer. Terms can get in your way because they are often ambiguous. Always focus first on the concepts, and then think about terms.

Classes, properties, and individuals are resources, and must have globally unique resource identifiers (IRIs). Ideally, your IRIs should be URLs, so people can access them on the Web. OWL does not make the unique name assumption. This means more than one IRI can refer to the same underlying resource. A namespace corresponds to a subset of IRIs managed by some governance body that mints them and ensures their uniqueness.

Reuse, Modularity, Triple Stores, and SPARQL

It is good practice to create ontologies in a modular way so that they can be reused by yourself or others. Ontologies are linked using an import mechanism.

Triples are used to represent OWL ontologies and all the data that populates those ontologies. They link together to form a knowledge graph. They are placed in a triple store and SPARQL is the language for querying the triples. Importantly, the triple store holds both the data and the metadata and both are queryable by SPARQL.

Although everything is represented as triples, it is important to separate out the ones representing the subject matter (metadata/TBox) from those expressing data (ABox) regarding that subject matter. Also, ontologies should be divided into separate modules, each representing a reusable slice of the subject matter at hand.

3.4   SUMMARY FOR PART 1

An ontology is a model of some subject matter that you care about. It specifies what kinds of things there are and how they are related to each other. OWL is a formal language for representing an ontology. An overview of OWL containing all of the key elements is depicted in Figure 3.7. The most important OWL constructs are `owl:Thing`, `owl:Class`, `rdf:type`, `rdfs:subClassOf`, `owl:ObjectProperty`, `owl:DatatypeProperty`, and `owl:Restriction`.

OWL supports automated reasoning, which makes the computer seem smarter and helps catch errors. The meaning of each OWL construct is understood in terms of the inferences that are sanctioned. Constructs can be combined in principled ways to form expressions whose meaning is derived from the meaning of the parts.

Sets are the foundation on which the meaning of every OWL construct is understood. Both the ontology and the data are represented in a single triple store. An IRI is a globally unique identifier which is a critically important part of OWL. The uniqueness enables reuse of ontologies and simplifies data integration. A namespace defines a set of IRIs, typically controlled by a single governance body.

The following explains the conventions used in Figure 3.7. The upper and lower sections in black apply across the board to OWL. The focus of the left column is individuals and literals—things that do not represent sets. The focus of the center column is classes and datatypes, which represent sets of individuals and literals, respectively. The right column pertains to properties.

The things that are lower down in the diagram are more foundational. As you go higher, you add information to or qualify things below. For example, individuals can be said to be the same as or different from each other. Hierarchies, disjointness, and equivalence pertain to both classes and properties. Properties are further qualified by domain and range and characteristics like transitivity. The upper portion shows the key ways one can build up expressions using things sitting lower down in the diagram.

This concludes Part 1 of this book. You have had a thorough introduction to what OWL is all about, and what it is founded on. You have seen the 30% of OWL that you will be using 90% of the time. If you have read and understood the material so far, you have a firm grasp of what OWL really means, and will be able to move forward with confidence in using it to build ontologies. Part 2 of the book expands and deepens your knowledge of the fundamentals covered in Part 1.

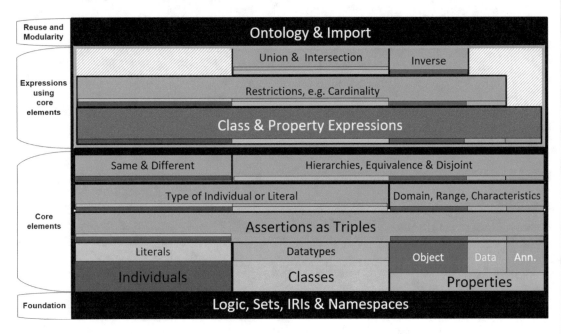

Figure 3.7: Overview of OWL.

Part 2

Going into Depth: Properties and Classes

Part 1 gave a thorough introduction to OWL touching on the main things that you will need when building ontologies. In this part of the book, we expand on the core elements.

CHAPTER 4

Properties

A property represents a way that two things can be related. Properties provide the glue that holds the Semantic Web together. This chapter gives more of a theoretical underpinning and explains how semantics can be given to properties. We will first describe object properties, which link individuals to other individuals, followed by data properties, which link individuals to literals.

4.1 PROPERTIES, RELATIONSHIPS, AND SETS

Consider the relationship between a subsidiary and its parent company. Alphabet Inc. has several subsidiaries, including Google and Nest Labs. Berkshire Hathaway subsidiaries include GEICO and The Buffalo News. Each of these relationships can be represented as a triple using an object property as follows:

```
:_Google     :isSubsidiaryOf :_Alphabet.
:_NestLabs   :isSubsidiaryOf :_Alphabet.
:_GEICO          :isSubsidiaryOf :_BerkshireHathaway.
:_TheBuffaloNews :isSubsidiaryOf :_BerkshireHathaway.
```

If we collect together all the triples that use this property, we get a set of ordered pairs of companies as follows: (Google, Alphabet), (Nest Labs, Alphabet), (GEICO, Berkshire Hathaway), (The Buffalo News, Berkshire Hathaway), and many others. *A set of ordered pairs*—that is what an OWL property is from a mathematical perspective. It is spelled out as such in the formal semantics of OWL.

An object property represents a *way* that two resources can be related to each other. Formally, an object property is the set of all ordered pairs of resources that are in fact related to each other in that particular way. In our example, that "way" corresponds to the first resource in the pair being a subsidiary of the second resource in the pair. This is the real-world semantics of the property `isSubsidiaryOf`. For brevity we use usually the term "pair," but we are always talking about ordered pairs.

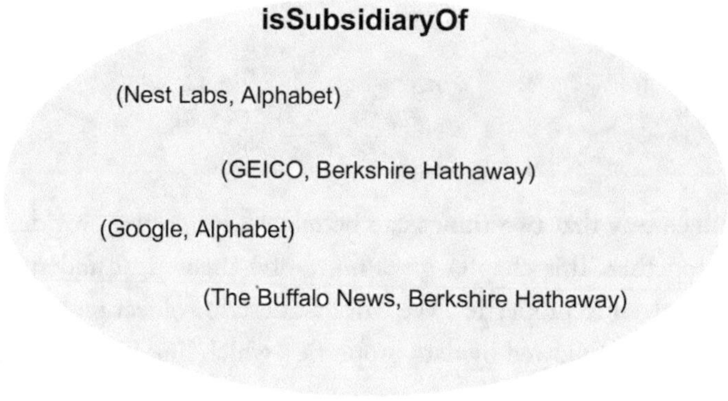

isSubsidiaryOf

(Nest Labs, Alphabet)

(GEICO, Berkshire Hathaway)

(Google, Alphabet)

(The Buffalo News, Berkshire Hathaway)

Figure 4.1: Properties as sets.

We saw some Venn diagrams in Chapter 2 where the sets were classes. A subclass is shown as one set being inside another. There is a direct parallel with subproperties. A super-property of `isSubsidiaryOf` would be a set of ordered pairs that included all the ordered pairs in the above diagram. A subproperty would include only a subset of the ordered pairs. Thinking of OWL properties in terms of sets can help disentangle meaning when things get a bit complex, especially when debugging.

Exercise 1: Can you think of examples of real-world properties that might be either a subproperty or superproperty of `isSubsidiaryOf`?

4.2   PROPERTIES ARE FIRST-CLASS OBJECTS

All modeling languages and paradigms have a way to represent a kind of thing. The term "entity" is frequently used for data modeling for relational databases. An entity results in a table; relationships and attributes result in columns. The term "class" is used for object-oriented modeling and programing and also in OWL—but, importantly, the term "class" does not have the same meaning in each context.

In general, kinds of things are characterized by what attributes and relationships their instances typically have. For example, companies have CEOs, subsidiaries, stock prices, and a number of employees. Employees have names, dates of birth, bosses, salaries, and organizations or departments that they work in.

An "entity" in entity-relationship modeling is much like a "class" in object-oriented modeling (hereafter an OO-class). In both cases, the relationships and attributes *belong* to the entities and OO-classes. When an instance of an OO-class or entity is created, the set of relationships and attributes of the class or entity determine the structure of the instances. An instance of an entity corresponds to a row in a database table. There is a column for each relationship or attribute, and NULL values are possible. When an object in object-oriented programming is instantiated, a data structure is created with space for each relationship or attribute. In this sense, entities and object-oriented classes act as *templates* for creating instances. There is space set aside whether it is used or not.

Because they belong to particular entities or OO-classes, attributes and relationships do not have independent existence. For example, if you want to model the subsidiary relationship between companies you might create a relationship called "parent" and it would belong to the class or entity called `Company`. It would become a column in a table, or a field in a data structure for an object.

If you wanted to model being a biological parent, which would be important for child support cases, you might have an entity or object called `Person` and use the same name "parent" for the biological parent relationship. Although there are two different relationships called "parent," each is subserviant to its OO-class or entity. This means there is no possibility of mixing up the two. If you are lucky, the modeler will document the meaning of the relationships, but often you just see the name and you make your best guess about its meaning.

In this case, it is a good thing that the two uses of parent cannot be confused, given their very different meanings. However, some relationships commonly occur for many kinds of things, e.g., being a part of something. A steering wheel is a part of a car. A chapter is part of a book. A screen is part of a laptop. In entity relationship (ER) and object modeling, because the relationships belong to their entities or classes, it is necessary to create several different versions of what is essentially the same relationship: being a part of something else. In OWL, it works very differently.

An OWL Property Stands Alone

In OWL, a property is a first-class object. It does not belong to any particular class, it can be used with any number of classes in a variety of ways. For example, in OWL we can create a property called `partOf` and use it to relate steering wheels to cars, chapters to books, and screens to laptops. The property is not dependent on any class for its existence, and is not owned by any class. We can create and use the inverse of the `partOf` property (`hasPart`) to say that an instance of a car has a part that is a steering wheel. Not all properties can be used in such a wide variety of contexts—e.g., `isSubsidiaryOf` only really makes sense between companies.

OWL Properties are Reusable

A major consequence of OWL properties being first-class objects is that they can be reused in a variety of situations.

An OWL Class is not a Template

Another consequence of properties being first-class objects is that OWL classes do not act as templates when instances are created. No space has to be allocated for NULLs in a relational table, or for empty fields on an object. Instead, space is allocated only when the actual triple is created.

We Can Specify the Semantics of a Property

A third major consequence of properties being first-class objects is that we can define and specify the meaning of a property in one place and reuse it with any number of classes. There are various things we can say about properties to specify their meaning. We saw examples of this in Section 2.3.5. The main ways to specify meaning for properties are:

- property hierarchies;

- domain and range;

- property characteristics (e.g., functional, transitive); and

- property inverses.

The next few sections of this chapter will explain how this works.

4.3   PROPERTY HIERARCHIES

Class hierarchies are very common, we saw one in healthcare in Figure 2.6. Because classes represent sets, we can illustrate classes and their subclasses using Venn diagrams.

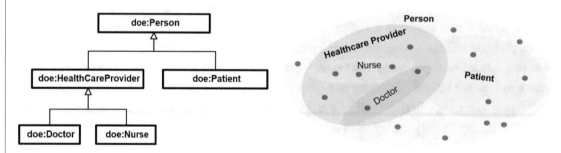

Figure 4.2: Class hierarchy as a Venn diagram.

None of the classes are known to be disjoint, so the diagram allows for the possibility that the same individual is a member of all three classes: Doctor, Nurse, and Patient.

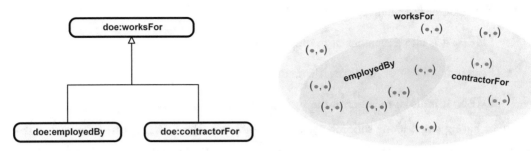

Figure 4.3: Property hierarchy as a Venn diagram.

But we just saw that properties also represent sets, so we can draw Venn diagrams for them too. The main difference is that whereas classes represent sets of *individual* resources, properties represent sets of *ordered pairs* of resources (see Figure 4.3).

The set perspective for properties illustrated in the Venn diagram makes it easy to see by inspection that anyone that is a *contractorFor* a given company also *worksFor* that same company. The same ordered pair of resources is in both sets. There are also pairs that are neither for employees or contractors. This allows for the possibility of other ways of working. Finally, there is one pair that is in the intersection of the two sub-property sets. This allows for the possibility of a given person being an employee and contract worker for the same company.

These things may or may not be true for your particular organization. To describe inferences, we will transition from speaking about ordered pairs to writing down triples. The triple, "x p y" is a shorthand for saying that the ordered pair(x, y) is a member of the set of pairs represented by property p.

Exercise 2: What would the Venn diagram in Figure 4.3 have to look like to rule out being an employee and a contractor for the same company?

Sanctioned Inferences

Specified as an inference rule, where the notation "x p y" indicates an asserted triple, the meaning of subproperty is:

IF: `p1 rdfs:subPropertyOf p2.` and
        `x  p1  y.`
THEN: `x  p2  y.`

So, when viewed from a set perspective, `rdfs:subClassOf` and `rdfs:subPropertyOf` are very close in meaning:

- IF: a resource is a member of the subclass

  THEN: it is also a member of the superclass.

- IF: an ordered pair of resources is a member of the subproperty

  THEN: it is also a member of the superproperty.

You cannot have a stand-alone subproperty or superproperty. The OWL construct, `rdfs:subPropertyOf` is a property representing a way that two properties can be related to each other.

Similarly, you cannot have a stand-alone subclass or superclass. The OWL construct, `rdfs:subClassOf` is a property representing a way that two classes can be related to each other.

4.4   DOMAIN AND RANGE

Creating subproperties is one way to specify meaning for a property; it does so by relating a property to another property. Another way to give meaning to properties is to specify when it makes sense to use a given property. For example, it makes no sense to say that a laptop has a subsidiary, nor could a biological parent be a corporation. Recall from Section 2.3.2 that "domain" translates informally to "only applies to." If a property has domain `C`, then it only applies to instances of `C`. Similarly, "range" translates to "range of possible values." If a property has range `C`, then the range of possible values that a property can have must all be instances of `C`.

More formally, consider the triple: `x  p  y`. The domain of property `p` answers the question: "What class is the subject necessarily a member of?" The range of `p` answers the question: "What class is the object necessarily a member of?"

Consider `isSubsidiaryOf`. If you know that `x` is a subsidiary of `y`, what can you say about `x` and `y`? A good place to start is to name the kinds of things you intend to use the property for; here the property will be used to link two companies. So the class `Company` is a candidate for both the domain and the range. However, for determining domain or range, focusing on what you merely *intend* is not going to work. For `Company` to be both the domain and range for `isSubsidiaryOf`, the subject and object must be instances of `Company` *in every conceivable use of the property*. It would be a logical necessity, much more than just desirable or intended.

These are strong criteria. However, I cannot think of what else `isSubsidiaryOf` might be used for and if someone wanted to use the property for something else, it would likely be wrong or an obscure edge case that you can safely ignore. So we are justified in using `Company` as the domain and range of `isSubsidiaryOf`.

Consider `partOf`. It is used to link a wide variety of things. That would include physical objects, countries, agreements, and books. In this case, it is probably best to not specify a domain or a range. Doing so could give wrong results.

For example, say the domain and range of `partOf` were specified to be a class called `Physicalobject`. Then someone decided to use it to model different parts of agreements. Then those agreements and their parts would be wrongly inferred to be physical objects.

These are two extreme examples. Most of the time, it is not so easy to decide. There is a strong tendency for ontologists to over-constrain domain and range. This limits property reuse; we will see other examples of this in the next section.

Sanctioned Inferences

IF: the triples: `p rdfs:domain C.` and `x p y.` are present
THEN: `x rdf:type C.` is inferred.

IF: the triples: `p rdfs:range C.` and `x p y.` are present
THEN: `y rdf:type C.` is inferred.

What happens if there is more than one domain or range, i.e., what happens if all three of the following triples are present?

```
p rdfs:domain C1.
p rdfs:domain C2.
x p y.
```

We can work this out from the sanctioned inferences above. The first domain allows us to infer that `x rdf:type C1`. The second domain allows us to infer that `x rdf:type C2`. Therefore, the net effect is for the domain to be the intersection of `C1` and `C2`. It would be logically equivalent to specify the domain using the following expression of the sort we discussed in Section 3.1.7.

```
p rdfs:domain (C1 and C2).
```

The situation is exactly analogous for range. Specifying two ranges is logically equivalent to specifying a single range to be the intersection of the two individual ranges.

4.4.1   USE DOMAIN AND RANGE WITH CARE
It's easy to make mistakes using domain and range. We consider several examples.

Multiple Domains or Ranges are Intersected

Suppose we have a property for borrowing. The borrower could be either a person or organization, and so can the lender. An easy mistake to make would be to do this:

```
borrowsFrom rdfs:domain Person.
```

```
borrowsFrom rdfs:domain Organization.
borrowsFrom rdfs:range  Person.
borrowsFrom rdfs:range  Organization.
```

Can you see the problem? This means that if we assert that
`_JohnDoe borrowsFrom _CommunityBank` then John Doe will also be inferred to be an organization (as well as a person) and the Community Bank will be inferred to be a person (as well as an organization).

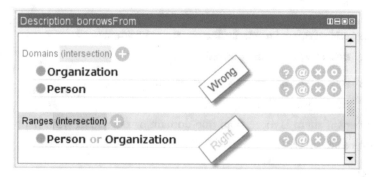

Figure 4.4: Domain and range in Protégé. Separate domain or range classes are intersected.

Instead the domain and range should be specified as the *union* of the two classes, `Person` and `Organization` (see Figure 4.4). Recall that Protégé is a commonly used ontology authoring and viewing tool. The sanctioned inferences would then correctly say that both John Doe and the Community Bank must each independently be *either* a person or an organization. Because this commonly arises, it makes sense to have a class that means: "either a person or an organization." That is exactly how `Party` is defined (see Figure 2.13). In Manchester syntax, this is:

```
Class: doe:Organization
Class: doe:Person
Class: doe:Party
    EquivalentTo:
        doe:Organization or doe:Person.
```

Don't Define Domain or Range Too Narrowly

Another common source of mistakes is to define domain and range too narrowly. For example, suppose you want to reuse the W3C media ontology.[16] A large number of properties use the class `MediaResource` as their domain. These include data properties such as `date` and `descrip-`

---

[16]   https://www.w3.org/TR/mediaont-10/.

tion, and object properties such as `hasContributor`, `hasLanguage`, and `hasPolicy`. This means anything that can conceivably have any of these properties is a `MediaResource`—by definition. Because no company, country or person can also be a `MediaResource`, the domains of the above properties dictate that a company cannot have a policy, a country cannot have a language and no person can contribute to manufacturing a car. This is dictated by their use of the OWL construct, `rdfs:domain`. There is nothing to stop you from asserting that a country has a language, but when you run inference the ontology will be inconsistent (we will see an example shortly).

Even if you are a media company whose bread and butter is exactly the production and sale of media resources, there are other uses of the properties that makes sense. If you decided to use this media ontology as part of your larger enterprise ontology, then it means you have to create duplicate properties for `hasLanguage`, `hasPolicy`, etc. that can be used in a context broader than media resources.

Let's take one of these and explore further: `hasLanguage`. What exactly can have a language? Media resources are the obvious answer, but a country can have an official language. A technical conference might have an official language. If you were 100% certain that these were the *only* things that could have a language, then you could create a class that was equivalent to the union of these classes and make that the domain. But that class is not very meaningful: "`Conference or MediaResource or Country`." In these cases, it can make sense to not have a domain at all. The range is a bit easier; it must be a `Language`.

Don't Regard Domain and Range as Integrity Constraints

A third major mistake is to regard domain and range as integrity constraints that are used by a gatekeeper to reject non-conformant data from entering a database. The formal semantics of domain and range have surprising consequences, especially for ER and OO modelers. Consider the `hasLanguage` property from the W3C media ontology. Suppose someone tries to assert the triple: `_Italy hasLanguage _Italian`. If domain acted as an integrity constraint, the assertion would be rejected. However, instead, the domain of `hasLanguage` sanctions the inference engine to infer the nonsense assertion that `_Italy` is an instance of `MediaResource`. The inference engine is doing the right thing, the model is wrong.

You can leverage the inference engine to find such mistakes, but you may need to give it more information. You need to say that Italy is a country and that countries and media resources are disjoint. That will cause the ontology to be inconsistent. Garbage-in, garbage-out applies here.

The user made what seems to be a perfectly sensible assertion, and the ontology was inconsistent. This is a trigger to go track down the error, which in this case was an overly narrow domain axiom. Here are the needed triples:

```
:_Italy rdf:type :Country.
```

```
:MediaResource owl:disjointWith :Country.
```

The use of domain and range in conjunction with disjoint axioms is powerful way to catch errors. It is important to include the disjoint axioms, and to say what classes individuals belong to. Otherwise, the logical errors lurking in the ontoloty could cause problems later on.

Summary: Domain and Range.

Given two things connected by a given property, domain and range are used to indicate what classes those two things necessarily belong to. Use with care. Don't think of them as integrity constraints. Specifying domain and range too narrowly is often worse than not specifying them at all. Be careful with multiple domains and ranges—it implies intersection.

4.5   INVERSE PROPERTIES AND PROPERTY CHAINS

Another way we can give a property meaning is to define it in terms of other properties. We will look at two examples: inverse properties and property chains.

4.5.1   INVERSE PROPERTIES

Each triple has a subject and an object, and the order is important. To say that `Google isSubsidiaryOf Alphabet` is taking the perspective of the smaller entity. If you are interested in the [opposite] perspective of the larger entity you might say `Alphabet hasSubsidiary Google` (see Figure 4.5).

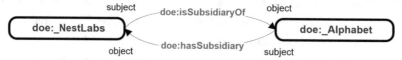

Figure 4.5: Property perspectives.

Taking the opposite perspective just means inverting the subject and the object. In terms of sets, you just invert the order of the items in the ordered pairs. Thus, the property from the opposite perspective is called the *inverse* of the original property (see Figure 4.6).

In English, taking the opposite perspective is just changing the way we write the sentence; the following two sentences are saying the same thing in two different ways.

1. "Nest Labs is a subsidiary of Alphabet" and

2. "Alphabet has Nest Labs as a subsidiary."

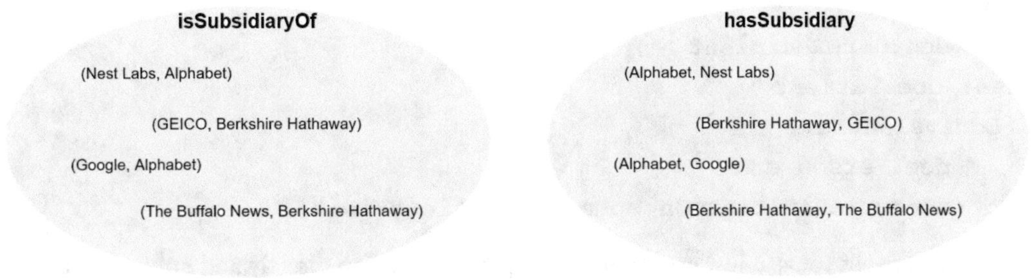

isSubsidiaryOf

(Nest Labs, Alphabet)

(GEICO, Berkshire Hathaway)

(Google, Alphabet)

(The Buffalo News, Berkshire Hathaway)

hasSubsidiary

(Alphabet, Nest Labs)

(Berkshire Hathaway, GEICO)

(Alphabet, Google)

(Berkshire Hathaway, The Buffalo News)

Figure 4.6: Inverse properties.

So in one sense there really is only one relationship, but in practice you have to choose a perspective to communicate which resource *is* the subsidiary vs. which resource *has* the subsidiary. It's a bit like a coin: it has two sides, but you can only look at one side at a time. It's also a bit like two quantumly entangled particles. If you change the spin of one, then the spin of the other changes simultaneously, even if it is millions of light years away.

Choosing a Perspective

When we speak of triples, choosing a perspective for a property means fixing the subject and the object. When we draw a triple, the perspective corresponds to the direction of the arrow (*from* the subject, *to* the object). See Figure 4.5.

When you create a property in OWL choose the perspective that seems the most natural way to think about it, or the way that it will most frequently arise when you populate the ontology and write SPARQL queries for intended usage scenarios. For example, consider a property that connects a loan contract to the borrower. It is central to the idea of a loan contract to know who the borrower is. However, from the perspective of a person or an organization, what loans they have is but one of many things that might be of interest. This suggests using the perspective where the loan contract is the subject and the borrower is the object. The property might be called "hasBorrower."

Recall also the example from Figure 2.10. There was the property careRecipient and its inverse, careRecipientOn. Which perspective seems most natural and intuitive to you? Note that there is not always a preferred perspective, both may be equally useful.

Using Inverse Properties

When inverses are needed you can do one of two things. First, you can create an explicit named property that is the inverse of an existing property and use it as you would use any other property. In Manchester syntax:

```
ObjectProperty: doe:careRecipientOn
```

```
InverseOf:
     doe:careRecipient
Class: doe:Patient
   EquivalentTo:
      doe:Person and
    (doe:careRecipientOn some doe:PatientVisit)
```

Alternatively, you can just use an object property expression directly and not bother to create an explicit inverse property.

```
Class: doe:Patient
   EquivalentTo:
      doe:Person and
    ((owl:inverseOf doe:careRecipient) some doe:PatientVisit)
```

The only difference between the two snippets of OWL is that the property used to define the restriction is named in the first snippet and is a property expression in the second. The property expression gives rise to an anonymous property just like class expressions give rise to anonymous classes, as we saw in Sections 2.3.4 and 2.3.5.

Explicit Inverse Properties: Yay or Nay?

It is not unusual to see ontologies where many if not all properties have explicit named inverses. However, there are limited upsides and several downsides.

1. If you store inferred triples, it will double the number of triples for that predicate.

2. It adds to the number of properties.

3. It's not always easy to find good names or definitions

Inferred triples: Most inferences say new things, but inverse property inferences just say the same thing in a different way. If you run inference and store the results in a triple store, it doubles the number of triples for a given property. So there is a cost with limited upside.

Extra properties: You can clutter up the ontology with unnecessary properties. Strictly speaking, explicit inverse properties are never necessary because you can always use an expression instead.

Names: When you do want the inverse, you must think of a name for it. The inverse of the property, isSubsidiaryOf, has an obvious name: hasSubsidiary, and hasPart is a good name for the inverse of partOf. But thinking of a name for the inverse that is reasonably natural English is not always easy.

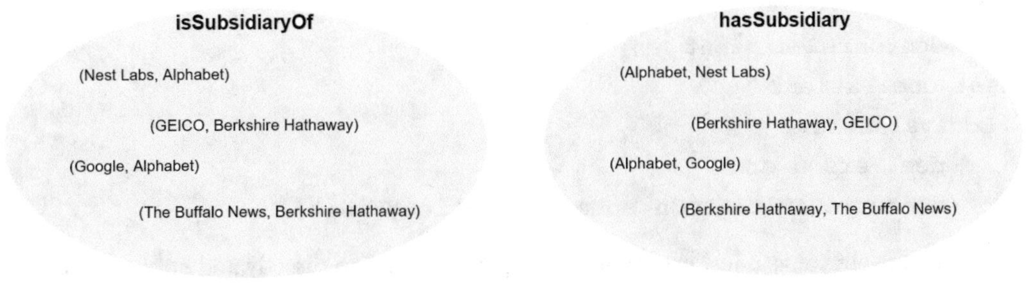

isSubsidiaryOf	hasSubsidiary
(Nest Labs, Alphabet)	(Alphabet, Nest Labs)
(GEICO, Berkshire Hathaway)	(Berkshire Hathaway, GEICO)
(Google, Alphabet)	(Alphabet, Google)
(The Buffalo News, Berkshire Hathaway)	(Berkshire Hathaway, The Buffalo News)

Exercise 3: Can you think of a good name for the inverse of `hasBorrower`? What about `works-For`? Sometimes there aren't any good names.

Here are some hints for the exercise. It is good practice to use names so that when used in a triple, it reads roughly like a sentence. For example, the triple:

`:_LoanContract_857466 :hasBorrower :_Person_JohnDoe`

is fairly close to the real sentence: "Loan contract 857466 has borrower: John Doe."

What about the triple going in the other direction? Try filling in the blank in this sentence: "John Doe _____ loan contract 857466." Then use that as the basis for deciding on a name of the inverse property that you would be happy to have in your ontology. What did you come up with? Do the same thing for the `worksFor` property by inverting the sentence: "Tim Cook works for Apple." Fill in the blank: "Apple _____ Tim Cook" then try to find a good name for the inverse property.

Even if you can find a good name, it may not be easy or particularly useful to come up with a good text definition. One way around this is to have the definition simply say: "the inverse of ..." where the ellipsis is replaced by the property name.

Sanctioned Inferences.

A property is the same as its inverse except that the subject and object are reversed. That's it. For example, if Scotland is `partOf` the UK then the UK `hasPart` Scotland. Also, if `hasPart` is the inverse of `partOf`, then it follows that `partOf` is the inverse of `hasPart`. So, the inferences warranted for inverse properties are quite straightforward:

IF: `p_inv owl:inverseOf p.` and `x p y.`
THEN: `y p_inv x.`

In addition,

IF: `p_inv owl:inverseOf p.`
THEN: `p owl:inverseOf p_inv.`

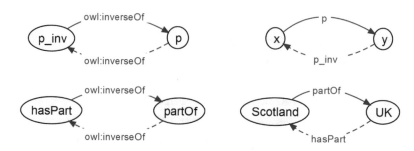

Figure 4.7: Inverse property inferences. Dotted lines indicate inferred triples.

Summary: Inverse Properties

When creating object properties, choose the perspective that is likely to be used most often and is most intuitive. The main time when you need to refer to an inverse property is for restrictions that require the opposite perspective from an already defined property (see Figure 2.10). When inverses are needed you can either create an explicit inverse property or you can just use an expression that refers to the existing property.

There are numerous downsides to explicit inverse properties, so create inverses only if you can see a clear benefit. The main justification for doing the latter is if property is routinely used in both directions, and the future users will expect to see both.

Please note: the recommendations and guidelines you see here are just that. There are good ontologists that I personally know and respect who take a different view about property inverses. They prefer to have them all or most of the time and to give them names and definitions. You will need to decide what works best for you. For additional discussion about inverse properties, see the blog article: "Named Property Inverses: Yay or Nay?"[17]

4.5.2   PROPERTY CHAINS

The second way to define a property in terms of other properties is by chaining two or more properties together. For example, if genealogy was important to your business, you might want to define an uncle to be the brother of a parent. In the example in Figure 4.8, you want to infer that Gene is Michael's uncle if you know Gene is the brother of one of Michael's parents. The figure uses the notation p1 o p2. The "o" symbol means composition; it is borrowed from mathematics. A new mathematical function can be composed by chaining one function after the other. Here we are doing something similar for OWL properties.

---

[17]   https://semanticarts.com/blog/named-property-inverses/.

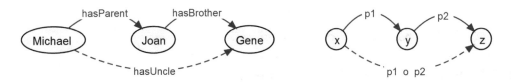

Figure 4.8: A property chain for uncle.

How It Works

The way OWL handles property chains is incomplete and slightly confusing. The obvious thing to do here would be exactly analogous to what we have done before with class expressions. We created the class expression (`Organization or Person`) and set it to be equivalent to the class `Party`. In effect this was giving that class expression a name. By analogy, we would like to give the property expression (`hasParent o hasBrother`) the name `hasUncle` by using `owl:equivalent-Property`. This would meet the main requirement to be able to infer that Michael has an Uncle Gene. However, for technical reasons beyond the scope of this book, it does not work that way.

Instead the inference requirement is met by making the property chain expression a *subproperty* of the new property, `hasUncle`. If the pair, (`Michael, Gene`) is in a subproperty of `hasUncle`, then (`Michael, Gene`) is also in the property `hasUncle`. Viewing properties as sets allows you to see this by inspection. This is depicted in Figure 4.9.

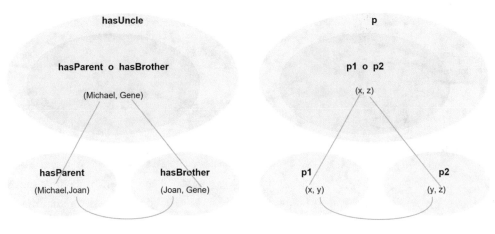

Figure 4.9: Venn diagrams for property chains.

In triples:

IF: :_Michael (hasParent o hasBrother) :_Gene. and
  (:hasParent o :hasBrother) rdfs:subPropertyOf :hasUncle.
THEN :_Michael   :hasUncle   :_Gene.

Another Example: Giving Care

Let's go back to our patient visit events (Figure 2.4) . A `PatientVisit` is linked to both a care provider and a care recipient. However, it might be of interest to answer the question: who are all the patients that Nurse Jane gave care to? There is no direct link. Furthermore, unlike for the uncle example, there is no path from `NurseJane` to either of the two patients, Ellen and John Doe, that goes in the direction of the arrows. To make a property chain that goes from a care provider to the patient, we must refer to the inverse of `careProvider`. We could use the expression (`owl:inverseOf careProvider`), or define an explicit inverse property and call it say, `careProviderOn`. They mean exactly the same thing (see Figures 4.10 and 4.11). There are now three cases where the two properties `careProviderOn` and `careRecipient` are chained together resulting in a third property called `gaveCareTo`.

Exercise 4: Draw a Venn diagram like the one in Figure 4.9 for this example. Hint: the property chain you want is: "`careProviderOn o careRecipient`." It will be a sub-property of `gaveCareTo`.

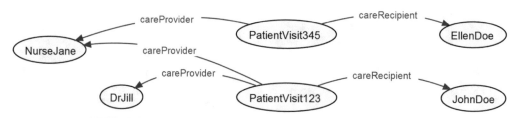

Figure 4.10: Patient visit events.

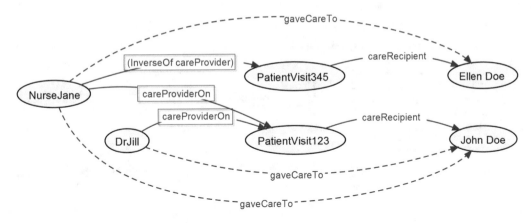

Figure 4.11: Property chain going through `PatientVisit`.

Practical Considerations

Tools provide different ways to create OWL property chains, and each syntax will show you something different. It is not clear from the syntax that property chains are implemented using subproperties. For example, here is what it looks like in Turtle:

```
doe:gaveCareTo rdf:type owl:ObjectProperty ;
   owl:propertyChainAxiom
      ( doe:careProviderOn
        doe:gaveCareTo
      ) .
```

Manchester syntax shows evidence of the idea of subproperty, but does not have an explicit subproperty relationship.

```
ObjectProperty: doe:gaveCareTo
   SubPropertyChain:
      doe:careProviderOn o doe:careRecipient
```

The syntax used in each case is shorthand for longer expressions using the formal subproperty axiom. It is more evident in the Protégé user interface.

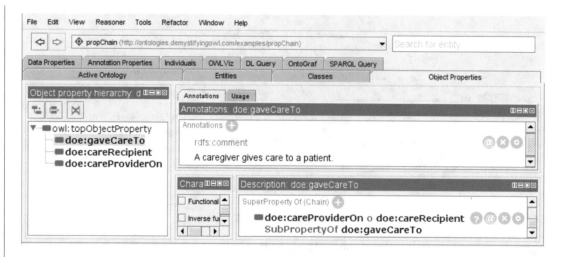

Figure 4.12: Property chains in Protégé.

To define a property chain in Protégé, for `doe:gaveCareTo`, you click the + button next to the text: "SuperProperty of (Chain)." The chain is defined using the `p1 o p2` notation. Protégé then tells you that the chain is a subproperty of `doe:gaveCareTo`.

Here, it helps to realize that if `rdfs:subPropertyOf` had an inverse, it would be called `rdfs:superPropertyOf`. It explains why you see both superproperty and subproperty in the Protégé window. One is from the perspective of the property chain itself, the other is from the perspective of the new property, `gaveCareTo`.

Sanctioned Inferences

IF: `p1 o p2.` is a subproperty of `p` and
          `x p1 y.` and `y p2 z.`
THEN: `x p z.`

Summary: Property Chains

Property chains are defined using the idea of function composition borrowed from mathematics. The desire for property chains arises frequently in modeling. However, there are some shortcomings that limit their use, the main one being that it cannot be used in a cardinality restriction. This is discussed more fully in Section 7.6. I have found the limitations often outweigh the benefits, and tend not to use them any more. If I need to use a property chain, I can always do it in SPARQL.

4.6   PROPERTY CHARACTERISTICS

We just looked at four ways we can specify the semantics of a property: by creating property hierarchies, specifying domain and range, being the inverse of an existing property and creating property chains. In this section we look at a fifth major way.

There is no limit to the number of ways two resources can be related to each other: having a biological mother, being older than, borrowing from, geographically bordering, being identified by, being a part of, among others. It turns out that if we look at large collections of properties, we notice that there are some characteristics that some properties have that others lack.

4.6.1   FUNCTIONAL PROPERTIES

For example, in Figure 2.6 we modeled the relationship between a patient visit and the care recipient so that there was never more than one care recipient. Properties with this characteristic are said to be *functional*. Another example is the relationship between various identifiers, and the things being identified. The whole point of an identifier is that it is identifying only one thing. A loan identification number identifies just one loan. A social security number identifies just one U.S. citizen. The two perspectives on this relationship are as follows.

1. The identifier *identifies* the thing.

2. The thing *is identified by* the identifier.

Can you think of a preferred perspective for this relationship? In this case, both are likely to be commonly used, and neither is obviously preferred. So let's call the two properties `identifies` and `isIdentifiedBy` and make one the inverse of the other.

The property `identifies` is functional. What about its inverse? Can a single individual have more than one identifier? Think about that for a few minutes. Consider the examples above from the perspective of the thing with the identifier: the employee, the U.S. citizen, the loan. For the property `identifiedBy` to be functional, it can only have one value for all and every conceivable scenario. You don't even need to be a citizen to have a national identification number. I lived in Scotland for many years and had a UK national insurance number that served much the same purpose as my U.S. social security number. So in fact, I was identified by two different numbers. Thus, `isIdentifiedBy` is not functional.

Suppose we preferred the perspective from the thing being identified so we created the property, `isIdentifiedBy`, but not its inverse. Since things can be identified by many things, this property is not functional. There is a way to say that the inverse of a property is functional without having to explicitly create the inverse and set it to be functional. We just say that the property `isIdentifiedBy` is *inverse functional*. Saying that a property has this characteristic is to say that its inverse property is functional (whether or not it is explicitly created).

If a given identifier is the subject of two triples using the `identifies` property and the objects have different IRIs, they are inferred to be the same thing. This is represented by the OWL construct: `owl:sameAs` (Figure 4.13).

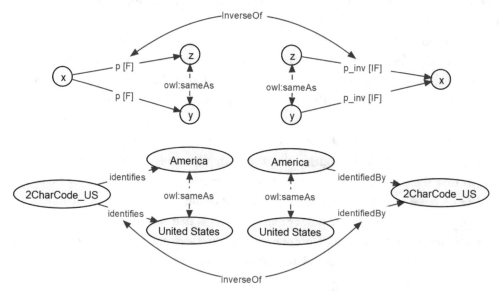

Figure 4.13: Functional [F] and inverse functional [IF] properties.

A functional property can only have one value for a given individual. The two-character ISO code that is "US" identifies only one country. If it points to two different URIs that means the two URIs refer to the same thing.

This is another illustration of how OWL does not make the unique name assumption. An inverse functional property is a property whose inverse property is functional. In the example, the functional property is `identifies`. Because the inverse of `identifiedBy` is functional, `identifiedBy` is by definition, inverse functional.

Other examples of properties that are functional in most situations include: `hasBoss`, `has-BiologicalMother`, and `hasCEO`. I say in most situations because it's not impossible to imagine having two people share a CEO role. With advances in medicine or freak situations in nature, someone could plausibly have more than one biological mother. Because you are building an enterprise ontology for practical purposes, it is usually safe to ignore such edge cases. When creating a property, always ask "Can this property ever have more than one value for a given individual?" If the answer is no, then the property is functional. If the answer is no for the inverse of a property, then the property is inverse functional.

Sanctioned Inferences.

Functional properties:

IF: `p rdf:type owl:FunctionalProperty.` and
         `x p y.` and `x p z.`
THEN: `z owl:sameAs y.`

Inverse functional properties:

IF: `p rdf:type owl:InverseFunctionalProperty.` and
         `y p x.` and `z p x.`
THEN: `z owl:sameAs y.`

4.6.2   TRANSITIVE PROPERTIES

Another characteristic that commonly arises, relates to what happens if you have a chain of triples linked by the same property. If you want to know who is richer or taller or faster than the next person you might have properties such as `richerThan`, `tallerThan`, and `fasterThan`. For each of these, if you chain them together the relationship will always hold between the first and last item in the chain, and everything in between. In general terms, these properties have the characteristic that if A is related to B and B is related to C, then A is related to C. This characteristic is called *transitivity*.

When you create a property, you should always ask: is it transitive? Some of OWL's built-in properties are transitive, for example: `rdfs:subClassOf`. If person is a subclass of primate and primate is a subclass of mammal, and mammal is a subclass of ... all the way up to lifeform, then person is a subclass of mammal and of everything in the chain up through lifeform (see Figure 4.14). Dotted lines depict sanctioned inferences.

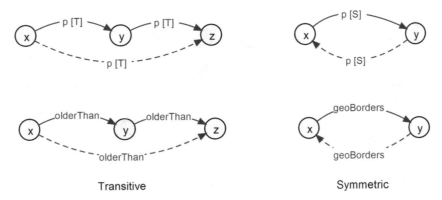

Figure 4.14: Transitive [T] and symmetric [S] properties.

Sanctioned Inferences

Transitive properties:

IF: `p rdf:type owl:TransitiveProperty.` and
       `x p y.` and `y p z.`
THEN: `x p z.`

4.6.3 SYMMETRIC AND ASYMMETRIC PROPERTIES

We have emphasized that order matters for properties. There is a real difference between being the parent or the child, or between being a subsidiary vs. having one. However, consider the relationship of sharing a geographic border. Saying that Canada borders the U.S. is no different in meaning than saying that the U.S. borders Canada. In this case, order does not matter. You can flip the subject and object and the relationship still holds. Another way of saying this is that *the property is its own inverse.* Properties with this characteristic are called symmetric. Another common example is `sib-lingOf`.

In the opposite situation, if the property holds in one direction, then it cannot possibly hold in the other direction. For example, if company A is a subsidiary of company B, then company B cannot possibly be a subsidiary of company A.

If a property is neither symmetric nor asymmetric, that means sometimes the relationship holds both ways and sometimes not.

Exercise 5: Think of a few more properties in each category: symmetric, asymmetric, and neither. Which category do you think will generally have the most properties? Why?

Sanctioned Inferences

Symmetric properties:

IF: `p rdf:type owl:SymmetricProperty.` and
       `x p y.`
THEN: `y p x.`

Asymmetric properties:

IF: `p rdf:type owl:AsymmetricProperty.` and
       `x p y.`
THEN: `¬(y p x).`

Recall that "¬" symbol is for logical negation, or "not" for short. It means what follows is provably false. You could think of this as negatively asserting a triple.

For Completeness: Reflexivity

There are two more property characteristics that I have never seen used in practice, but I mention them here for completeness: reflexive and irreflexive. Check out the OWL 2 documentation if you are curious.[18]

4.7 PROPERTY CHARACTERISTICS OF SUBPROPERTIES AND INVERSE PROPERTIES

4.7.1 SUBPROPERTIES

Next we consider what if anything can be inferred about the characteristics of a property given the characteristics of its super-property. For example:

IF: property p1 is functional, transitive and symmetric and

property p2 is a subproperty of property p1

THEN: what can we say about the characteristics of property p2?

Let's consider functionality first. This is a good time to consider properties from a set perspective. Stare at the Venn diagram in Figure 4.3. Recall that a property is a set of ordered pairs, and a subproperty is a subset of the pairs of the superproperty. For the sake of this discussion, let's say that `worksFor` is functional (i.e., you can only work for one person or organization). This means that no two pairs in the `worksFor` property can have the same subject but a different object. If you look at this in the right way, it becomes immediately obvious whether or not `employedBy` and `contractorFor` must also be functional. Can you see it?

Pick any two pairs from the `employedBy` subproperty. They are both in the `worksFor` property, and we already know that no two of those pairs can have the same subject and two different objects. Therefore, the subproperty of a functional property is functional. The same line of reasoning says that the subproperty of an inverse functional property is inverse functional.

You may be thinking that all property characteristics are in effect inherited from their super-property. Let's consider symmetry. The borders property is symmetric, if Canada borders the USA, then the USA borders Canada and vice versa. We can create a subproperty of borders for the special case that one country borders the other from the North. Now, Canada borders the USA from the North but not vice versa. Therefore,

- the subproperty of a symmetric property need not be symmetric.

What about transitivity? Again, it helps to think about the property as a set of pairs. Let P and subP be properties with the following pairs:

---

18 https://www.w3.org/TR/owl2-syntax/#Reflexive_Object_Properties.

- P = {(x,y), (y,z), (x,z)}

- subP = {(x,y), (y,z)}

The property subP is clearly a subproperty of P and not transitive, whereas the P is transitive. Therefore:

- the subproperty of a transitive property need not be transitive.

Note that the set perspective of properties sometimes allows you to more or less "see" the solution by merely inspecting Venn diagrams of subproperty relationships. The alternative way to convince yourself (or others) amounts to working out (often simple) step by step lines of reasoning which lead you to the conclusion you are after. This entails writing down expressions with triples. In the next exercise you will get a chance to do exactly that.

4.7.2 INVERSE PROPERTIES

Exercise 6: Answer the following three additional questions asking what we can know about the characteristics of the inverse of a property from the characteristics of the property.

1. If property **p** is functional, what can we say about whether **p_inv** is or is not functional?

2. If property **p** is symmetric, what can we say about whether **p_inv** is or is not symmetric?

3. If property **p** is transitive, what can we say about whether **p_inv** is or is not transitive?

Hint: it may help to draw some Venn diagrams representing properties as sets of pairs.

4.8 DATA PROPERTIES

A data property connects an individual to a literal rather than to another individual. Despite this important difference, data properties share much in common with object properties. They are first-class objects, they can be arranged in a sub-property hierarchy, they have domains and ranges, they can be functional, and they can be used to create property restrictions. We consider these in turn and highlight the differences.

4.8.1 DATA VS. OBJECT PROPERTIES

Recall from Section 3.1.5 that from a set perspective, datatypes are just like classes, except that their members are literals instead of individuals. Despite the close analogy between classes/individuals and datatypes/literals, there is a stark divide between individuals and literals that is enforced by OWL. We will not explore the technical reasons; instead, we focus on the ramifications.

The most direct consequence is that the range of every data property must be a datatype (hence the name). We saw examples of this already with the properties for names having a range of `xsd:string`, and the date of a patient visit having range of `xsd:dateTime`.

The fact that a literal cannot be the subject of a triple means that *a data property cannot have an inverse*. For example, given that `firstName` is a data property, we can say that someone's first name is "Jane," but OWL does not allow us to assert in a triple that the string "Jane" is the first name of someone.

This has major consequences on what characteristics make sense for datatypes. Before you continue reading, ask yourself: can a data property be inverse functional or transitive or symmetric? Why or why not? Refer to at Figures 4.13 and 4.14.

For a data property to be inverse functional, its inverse property has to be functional. But a datatype cannot have an inverse, so this characteristic cannot apply.

To be symmetric, a property has to hold in both directions. But a data property only goes in one direction, so a data property can never be symmetric.

To be transitive, a datatype has to be chained, meaning that the object of the first triple becomes the subject of the next triple. But a literal can never be the subject of a triple, so a datatype can never be transitive either.

What about functional? Suppose we want to say that a no one can have more than one last name. Does that make sense? It does. So being functional applies equally to object and data properties. It is the only property characteristic that applies to data properties.

4.8.2   WHEN TO USE DATA PROPERTIES

Often, you will create object properties because you know you are connecting to other individuals, so there is no question. Frequently, you will be inclined to create a data property, because it seems like the right thing to do. For example, in Chapter 2 we modeled first and last names as data properties. However, the right thing to do depends on what the intended usage of that property is. If you are tracking patient names, a data property is probably fine.

However, suppose your use case involves tracking large numbers of attorney names in a database. You could instead make `firstName` and `lastName` object properties both of whose ranges are a class called `Name`. In turn, each instance of the class `Name`, would be linked via a data property to the string representing the name. The data property might be called `text`. This means having individuals rather than literals that directly correspond to specific first and last names, e.g., `_Name_Johnnie` and `_Name_Cochran` (see Figure 4.15).

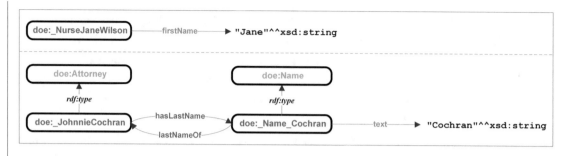

Figure 4.15: Names: datatype or object property?

Using an object property enables you to say something about the name. You could create an inverse of `hasLastName` that relates a name to a person with that name as their last name. That inverse property could be used to discover all the attorneys with a given last name. You could say other things about the name, e.g., which countries is it commonly used in. None of that is possible using a data property. At least not with the inference engine.

However, you can use SPARQL to find out who has a particular last name. The following returns all the (IRIs for) people whose last name is "Cochran" along with their first name.

```
SELECT ?person    ?firstName
WHERE    {?person rdf:type doe:Person.
          ?person doe:hasLastName "Cochran".
          ?person doe:firstName ?firstname.}
```

The effect is similar to traversing the arc from literal to individual. You start with the string "Cochran" and see what it is connected to via the `hasLastName` data property. Recall from Section 3.1.12 that there are two steps to executing a SPARQL query. First you find triples in the triple store that match the pattern in the WHERE clause. Once the individual with the specific last name is found, the first name is retrieved. Then a table of results is returned based on what is asked for in the SELECT part of the query. There could be a performance hit because of the string matching.

Another example is in the context of offers that must indicate the amount of something that is for sale. If you are selling a 50-lb bag of sand in your local classified ads, it's probably fine to have it all in a string: "50-lb of sand." The financial markets that are offering 50 million barrels of Brent crude oil will want to track things more carefully. For that, it will be necessary to separate out the product being sold (Brent crude) from how much of it is on offer (50 million barrels). To aggregate sales volumes where things are sold in different units will require separating out units of measure from the number amount. This is shown in detail in Figure 4.16.

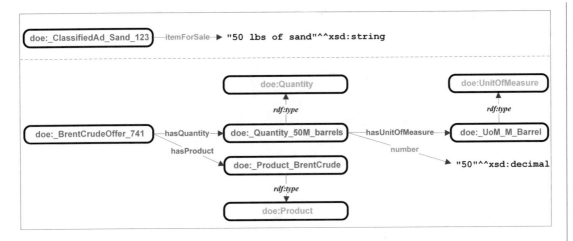

Figure 4.16: Amounts: datatype or object property?

Sometimes it is not so obvious what the best thing to do is. In that case, it boils down to answering the following question:

Given a property that has some kind of value, do you now or might you in the future have the need to say something about that value?

If so, then the value will have to be the subject of a triple and an object property will be required. If not, you can safely choose to use a data property. If you are uncertain, you could choose to future-proof your ontology by using object properties, in case you needed them later. The cost of doing this is the need to create an extra individual and extra triple each time you assign a value (as depicted in Figure 4.15).

4.9 DISJOINTNESS AND EQUIVALENCE

Figure 4.3 allows for a given individual to be both an employee and a contract worker for the same company. If you know this to be impossible, the two properties can be specified to be disjoint. Although we can think of a few examples, this is an infrequently used construct in OWL.

Exercise 7: Can you think of another example where two properties will be disjoint? Hint: think about loans, borrowers, and lenders.

Two properties can also be specified to be equivalent. In terms of sets, that means both properties have the exact same set of ordered pairs. The main situation when this is useful is if you are mapping one ontology to another and you need to keep both names around. For example, em-

`ployedBy` might be called `isEmployedBy` in another ontology. In Section 8.2.3, we will see an example of using equivalent property for mapping.

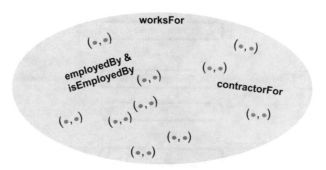

Figure 4.17: Disjoint and equivalent properties. Here, because they overlap exactly, `employedBy` and `isEmployedBy` are equivalent. Both are disjoint from `contractorFor`.

Sanctioned Inferences

Disjoint properties: If two properties are disjoint, then no single pair can be in both properties. In terms of triples:

IF: `p1 owl:propertyDisjointWith p2.` and
        `x p1 y.`
THEN: ¬`(x p2 y).`

Equivalent properties: If two properties are equivalent, then if a pair is in one of the properties, then it is also in the other. In terms of triples:

IF: `p1 owl:equivalentProperty p2.`
THEN: `x p1 y => x p2 y.` and
        `x p2 y => x p1 y.`

This has broad implications. If two properties are equivalent, then everything that is true about one is also true about the other. That applies to all the characteristics such as transitivity, functionality, symmetry etc.

Recall from the Section 3.1.1 that the "=>" symbol means logical implication (not greater than or equal to). It is a more compact way of saying IF <some premise> THEN <some conclusion>. Here it avoids ugly nesting of IF/THEN statements. Based on what subproperty means, we can rewrite the definition of property equivalence this way:

IF: `p1 owl:equivalentProperty p2.`
THEN: `p1 rdfs:subPropertyOf p2.` and
        `p2 rdfs:subPropertyOf p1.`

So, if two properties are sub properties of each other that means the two properties are equivalent to each other. There are two main situations when you use property equivalence:

1. when creating a crosswalk or mapping from one ontology to another and

2. when giving a property a new name and deprecating the old name. Deprecation is discussed in Section 8.4.5.

Exercise 8: A symmetric property is its own inverse. What similarly pithy thing can be said that characterizes the essence of what it means to be an asymmetric property? Hint, it involves disjointness.

4.10 ANNOTATION PROPERTIES

We have just completed a thorough treatment of object and data properties, addressing everything that you are likely to need in of your day-to-day OWL modeling. We looked at what you can say to define the meaning of the properties, and what that means in terms of what inferences are sanctioned.

There is one more kind of property that we briefly introduced in Chapter 2 that allows you to assign human readable labels and text definitions to classes and properties. These are examples of annotations that help humans understand the ontology and are accessible to SPARQL, but are ignored by the inference engine. There are nine that are provided in the OWL spec.[19] The most common ones are:

- `rdfs:label` attaches a human readable label to an IRI

- `rdfs:comment` attaches a human readable comment to an IRI.

Labels are often used by tooling. For example, the property `isSubsidiaryOf` might have the label "is subsidiary of" and the class `PatientVisit` might have the label: "patient visit." Comments can be used for anything, often they are definitions. For example, `PatientVisit` might have the comment: "The event of a person receiving healthcare."

---

[19] https://www.w3.org/TR/owl2-syntax/#Annotation_Properties or https://www.w3.org/TR/owl2-quick-reference/#Annotations.

Annotation vs. Object and Data Properties

Like object and data properties, annotation properties are used to assert triples, can have domain and range, and can be arranged in a subproperty hierarchy. Here is a triple assigning a label: `doe:PatientVisit rdfs:label "patient visit."`

The main difference is that annotations do not participate in inference. Another important difference is that there is no object vs. datatype distinction. The value of an annotation property can be either an IRI or a literal. For example, the following triple says that a particular class is defined in the Good Relations[20] ontology about product offerings on the Web:

```
gr:Offering rdfs:isDefinedBy
                <http://purl.org/goodrelations/v1>
```

where the `gr` is the abbreviation for the namespace prefix:

```
            http://purl.org/goodrelations/v1#
```

Using Annotations

Unless you have a good reason not to, you should generally use labels. Some tools will create them for you automatically converting from camel case. Many ontologies use `rdfs:comment` for a variety of things including offhand remarks hinting at the meaning, a carefully crafted definition, examples, counterexamples, notes on usage, and even notes about the development of the ontology itself.

It's a good idea to make it clear what kind of comment you are writing. A simple approach is to prefix each `rdfs:comment` with something like "DEFINITION:", "NOTE:", or "EXAMPLE:". An alternative is to create or use existing annotation properties for each of the separate purposes. For example, it is increasingly common for ontology authors to make use of `skos:definition`[21] for definitions. It is a matter of style and preference. Whichever approach you use, it is important to use it consistently.

Think about the consequences of creating a large number of different kinds of annotations. It means every time someone (including yourself) wants to add a comment, they have to scan through a large number of options to make sure they use the right one. It is also error-prone because people are often lazy and may not use the annotation properties consistently.

Before creating your own, look at the SKOS ontology for some commonly used annotations (see Figure 4.18). Notice how they are arranged in a hierarchy. There are three kinds of labels, and several kinds of notes. If you want a special annotation for synonyms or abbreviations, then you can create them and put them under `skos:altLabel` in the annotation property hierarchy.

---

[20] http://purl.org/goodrelations/v1.
[21] See: https://www.w3.org/2009/08/skos-reference/skos.html.

Figure: 4.18: Annotation property hierarchy.

Language Tags

If you anticipate creating multi-lingual applications, you can specify what language you are using for text annotations. For example, the English word "contract" is "Fabrik" in German and "contrato" in Spanish. If there was a class for contract in your finance ontology whose IRI was `fin:Contract`, you could assign it all three labels as shown below. The two character country codes (en, de, and es) are ISO standards.[22]

```
fin:Contract rdfs:label "contract"@en.

fin:Contract rdfs:label "Fabrik"@de.

fin:Contract rdfs:label "contrato"@es.
```

Machine Processable

Although annotations are ignored by inference engines, they are machine processable. For example, you could use SPARQL to find all the comments that have a particular string in them. There are commercial tools using natural language processing that read and processes the comments and labels for a variety of purposes. For example, some tools will do auto-tagging of text documents using the concepts in an ontology as tags. They analyze the comments to find the best match.

[22] https://en.wikipedia.org/wiki/ISO_3166-1_alpha-2.

Summary: Annotations

1. Annotation properties are much like datatype and object properties. The main difference is that they do not participate in inference.

2. The most commonly used annotations are for labels and comments.

3. Annotations are a good way to create synonyms and abbreviations

4. Annotations with literals as values are typically strings, but they can be of any datatype.

5. Like object properties, annotations can also have IRIs as values.

6. It is common practice to use SKOS annotations for labels, definitions, examples, and other notes.

4.11 SUMMARY LEARNING

Properties, Relationships, and Sets

OWL properties are used to represent real-world relationships. They are the glue that holds the Semantic Web together. Properties are stand-alone first-class objects that do not belong to any class and are broadly reusable. Underneath the covers, a property is a set of ordered pairs, just like classes are sets of individuals. Both subclasses and subproperties correspond to subsets.

Property Semantics

The semantics for a property can be specified in a number of ways. The subproperty relationship gives rise to a property hierarchy in the same way that the subclass relationship gives rise to a class hierarchy.

Special care must be taken when using domain and range, which indicate the kinds of things that are related. Don't think of them as integrity constraints. Don't specify them too narrowly, it limits reuse. Also, multiple domains or ranges implies intersection.

Properties can have one or more of a variety of characteristics, such as being functional, symmetric or transitive. A functional property can have only one value, a symmetric property is its own inverse and a transitive property allows inference along a chain of uses of the same property. Property chains use other properties to specify the semantics of a given property.

Inverse Properties

Order matters for properties. If you flip the subject and object, the property typically will not hold. Being a subsidiary is different from having a subsidiary. By flipping the subject and object, the inverse of a property is representing the same relationship from the opposite perspective. You can create explicit inverse properties, or you can use an expression to refer to a property's inverse. Create explicit inverses only when you need them. There are some downsides.

Data Properties

A data property is just like an object property, except that it connects an individual to a literal, not to another individual. Data properties cannot have inverses. Sometimes it is not clear whether to use a data property or object property. If you think you might need to say something about the value, then create an object property because a literal cannot be the subject of a triple.

Disjointness and Equivalence

Properties are sets, so they can be disjoint or equivalent, just like classes can.

Annotation Properties

Annotation properties are used to say things that do not participate in inference, but are accessible to SPARQL. Annotations can have literals or IRIs as values. The most commonly used annotations provided by OWL are `rdfs:label` and `rdfs:comment`. However it is becoming standard practice to use SKOS annotations instead.

CHAPTER 5

Classes

In this chapter we review the core concepts about classes that we saw in Chapter 2 and then explore them in more depth.

5.1   REVIEW: CLASSES AND SETS

OWL classes are used to represent different kinds of things such as person and corporation. Mathematically, a class corresponds to a set. All the members of the set share certain properties, which is why they are grouped together into a class. Classes can be arranged into hierarchies using `rdfs:subClassOf`. The subclass relationship means subset, and it corresponds to logical implication. For example, to say that `Nurse` is a subclass of `Person` is precisely to say that:

```
IF:     x rdf:type :Nurse.
THEN: x rdf:type :Person.
```

Finally, Venn diagrams are a helpful way to visualize classes and subclasses (see Figure 4.2).

5.2   CLASS RELATIONSHIPS

There are three main ways that classes can be related to each other, and all correspond to relationships between sets.

1. `C1 rdfs:subClassOf C2`

2. `C1 owl:equivalentClass C2`

3. `C1 owl:disjointWith C2`

5.2.1   SUBCLASS

Subclassing is used to indicate that one class is more specific than another. To say that `C1 rdfs:subClassOf C2` is precisely to say that for any individual, `x`

```
IF:     x rdf:type C1.
THEN: x rdf:type C2.
```

For example, to say that `Corporation rdfs:subClassOf LegalEntity` is to say that for any individual, `x`

```
IF:     x rdf:type :Corporation.
THEN: x rdf:type :LegalEntity.
```

This is depicted in Figure 3.1.

5.2.2 CLASS EQUIVALENCE

Class equivalence is used to state that the members of one class are exactly the same as the members of another class. To say that `C1 owl:equivalentClass C2` is precisely to say the following two things:

```
IF:     x rdf:type C1.          IF:     x rdf:type C2.
THEN: x rdf:type C2.          THEN: x rdf:type C1.
```

Note that the statement on the left is just what we saw above; it means `C1` is a subclass of `C2`. On the right, the two classes have been swapped, which therefore is the same as saying that `C2` is a subclass of `C1`. Thus to say that two classes are equivalent is really just a shorthand way to say that two classes are subclasses of each other. At the end of Section 4.9 we saw the same thing with equivalent properties and subproperty.

If we visualize this in terms of a Venn diagram where the classes are sets, `C1` subclass of `C2` means that `C1` is inside `C2`, and `C2` subclass of `C1` means `C2` is inside `C1`. The only way both can be "inside" each other is by having them be identical to each other.

This is related to the idea of proper subset, which means there are individuals that are not in the subset that are in the superset. Using the idea of subclass to define class equivalence tells us that `rdfs:subClassOf` does not imply proper subset, the sets can be the same.

There are two common uses of class equivalence. The main one is to specify additional meaning using class expressions. An example of this is depicted in Figure 2.9. This is explored in depth in Section 5.3 below. The other use is for stating that two classes from two different ontologies mean the same thing. The latter is not part of building an ontology, but rather of mapping one ontology to another.

5.2.3 DISJOINT CLASSES

Disjointness between classes is used to say that the classes cannot have any individuals in common. To say that `C1 owl:disjointWith C2` is to say the following two things:

```
IF:     x rdf:type C1.          IF:     x rdf:type C2.
THEN: ¬(x rdf:type C2).        THEN: ¬(x rdf:type C1).
```

Recall that "¬" is the symbol for logical negation—it means what follows cannot be true. The term "disjoint" comes from sets that do not share any members.

There are two main uses of class disjointness. One is to communicate to humans more about the meaning of the classes involved. An equally, if not more important, use is to help the inference engine spot logical errors in the ontology. If any individual is asserted to be a member of both classes, the inference engine will detect a logical inconsistency. We saw the pattern using high-level disjoint classes at the very end of Section 2.3.5 and we will see another example in Section 8.2.2.

5.3 CLASS EXPRESSIONS

As discussed in Section 3.1.7, we are familiar with the idea of building up expressions using operators. For example, `4*(11-3)` is an expression that indirectly refers to the number thirty-two. The simple expression, "32" is the direct way to refer to the same number. Similarly, a class expression in OWL is an expression consisting of classes and operators that indirectly refers to a class. In Section 2.3, we saw some examples:

1. `doe:Organization` or `doe:Person`

2. `doe:TwoWheeledVehicle` and `doe:MotorizedVehicle`

3. `doe:Person` and (`doe:careRecipientOn` some `doe:PatientVisit`)

These expressions indirectly refer to classes. Simple expressions that directly refer to classes are plain IRIs such as `doe:Person`, `doe:PatientVisit`, and `doe:Corporation`. They function more as names for classes.

Table 5.1: Expressions in arithmetic and in OWL

	Arithmetic	OWL Classes
Simple Expressions (direct reference)	An Arabic numeral directly refers to a number, e.g., 3, 32	An IRI directly refers to a class, e.g., `doe:Person`
Complex Expressions (indirect reference)	An expression indirectly refers to a number. `4*(11-3)` means 32. It is derived from the meaning of the operators and the numbers that make up the expression.	An expression indirectly refers to a class. (`TwoWheeledVehicle` and `MotorizedVehicle`) means motorcycle. Its meaning is derived from the classes and operators that make up the expression.

Just like for arithmetic expressions, the meaning of the expression is derived from the meaning of the parts of the expression. This gives us the ability to encode meaning of a class in terms of other classes, properties and occasionally, individuals. Among the most commonly used class operators are the OWL constructs: `owl:intersectionOf`, `owl:unionOf`, and `owl:someValuesFrom`.

In Manchester syntax these constructs show up as "and," "or," and "some," respectively. Further-more, the meaning of the constructs is available to an inference engine. An IRI by itself, with the addition of some text annotations, allows a human to get a good idea of the intended meaning. However, the inference engine has no access to that meaning.

Everywhere a class is expected in OWL, any class expression can be used (see Table 5.1).

5.3.1   ANONYMOUS CLASSES AND BLANK NODES

Each of the above class expressions describes a class, but they don't have IRIs or names and are thus called *anonymous classes*. For each of the above examples, we can think of good names and assign them to the classes, e.g., `Party`, `Motorcycle`, and `Patient`. We use `owl:equivalentClass` to do that (as we did in Section 2.3).

We want to have a triple with `Motorcycle` as the subject and `owl:equivalentClass` as the predicate. Somehow we want to achieve the effect of having an expression with no IRI be the object. Until now, objects of triples have always been either IRIs or literal values. For this, there is a third kind of node, which is used to represent anonymous classes. It is called a *blank node*. In the Turtle syntax, square brackets enclose a blank node (see below).

```
doe:Motorcycle
        rdf:type owl:Class ;
        owl:equivalentClass
            [ rdf:type owl:Class ;
              owl:intersectionOf
                    ( doe:MotorizedVehicle
                      doe:TwoWheeledVehicle
                    )
            ] .
```

A slightly simplified view of the triples are depicted in Figure 5.1. Make sure you can find in the Turtle each of the four triples that are in the figure. Blank node 1 is enclosed in square brackets in the Turtle. Blank node 11 shows up enclosed in round parentheses. The node in the lower right is dotted to indicate there are some missing details that are not important for now.

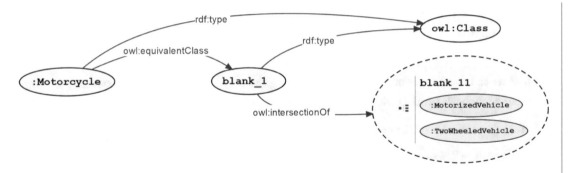

Figure 5.1: A simplified view of a blank node representing an intersection expression.

5.3.2   BOOLEAN EXPRESSIONS

Intersection

In Figure 2.12 we defined `Motorcycle` to be the intersection of `TwoWheeledVehicle` and `MotorizedVehicle`. This involved three steps.

1. Create a class called `Motorcycle` (using a simple class expression).

2. Create an expression that intersects the two classes.

3. Set `Motorcycle` to be equivalent to the intersection expression.

Sanctioned Inferences

In terms of sets, the intersection includes just those individuals that are members of both sets. Therefore, the inferences that are justified using the intersection construct are as follows.

IF: `C1 owl:equivalentClass (C2 and C3).`
THEN:

    1. `C1 rdfs:subClassOf C2.`
    2. `C1 rdfs:subClassOf C3.`

From this, and from the definition of `rdfs:subClassOf`,

IF: `x rdf:type C1.`
THEN:

    1. `x rdf:type C2.`
    2. `x rdf:type C3.`

Convince yourself that these inferences are justified. You may find it helpful to draw a Venn diagram.

Syntax Variations

So that we can focus more on ease of understanding the meaning and less on the syntax, I am using a mix of Manchester and Turtle syntax. The expression:

```
C1 owl:equivalentClass (C2 and C3).
```

is represented in Turtle as follows:

```
C1 rdf:type owl:Class ;
    owl:equivalentClass
        [ rdf:type owl:Class ;
          owl:intersectionOf ( C1 C2 )
        ] .
```

It looks like this, in Manchester syntax:

```
Class: doe:C1
    EquivalentTo:
        doe:C2
          and doe:C3
```

Note that in the Turtle above it explicitly says that `C1` is an instance of `owl:Class`. However, "`rdf:type owl:Class`" also appears directly after the open square bracket "`[`." That makes explicit that the blank node is also an instance of `owl:Class`, as depicted in Figure 5.1.

Manchester syntax is widely used in tools for creating class expressions. For most other things, people tend to use Turtle or RDF/XML, with use of the former becoming much more widespread and the latter receding in popularity.

Union

In Figure 2.13 we defined `Party` to be the union of `Person` and `Organization`. This also involved three steps.

1. Create the class `Party`.

2. Create the expression that is the union of `Person` and `Organization`.

3. Set `Party` to be equivalent to the expression (`Party or Organization`).

Sanctioned Inferences

In terms of sets, the union includes just those individuals that are members of either or both sets. Therefore, the following inferences using the union construct are justified.

IF: `C1 owl:equivalentClass (C2 or C3)`
THEN:
1.   `C2 rdfs:subClassOf C1.`
2.   `C3 rdfs:subClassOf C1.`

From 1 and 2 above, and from the definition of `rdfs:subClassOf`, we can conclude that:

IF: *either or both* of the two following triples exist:
1.   `x rdf:type C2 .`
2.   `x rdf:type C3.`
THEN:
3.   `x rdf:type C1.`

   There is one additional inference that is a bit more subtle. We have not yet introduced the OWL constructs needed to represent it in triples, but it works in the following way. If `C1` is the union of `C2` and `C3` as above, and if it can be proven that the triple `x rdf:type C2` is false, then the triple `x rdf:type C3` is inferred to be true (and vice versa, swapping `C2` and `C3`).

   For example, if you know that `x` is a `Party` which is defined to be the union of `Person` and `Organization`, and you also know that `x` is not a `Person`, then `x` must be an `Organization`. The Turtle syntax is exactly analogous to that for the intersection example with `intersectionOf` replaced by `unionOf`.

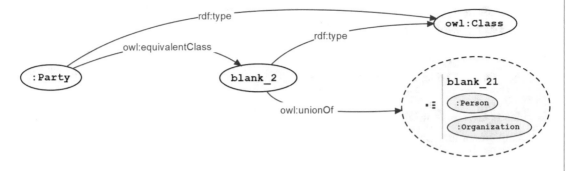

Figure 5.2: A simplified view of a blank node representing a union expression.

Complement

The final Boolean expression we discuss corresponds to the set complement operation. The complement of a set is defined to be everything that is *not* in that set. For example, the complement of the set of all organizations, is everything that is *not* an organization. This includes not only things that are relevant to organizations—such as people, agreements and contracts—but it also includes elephants, drugs, jokes, and pickles—and literally anything else that you can think of (besides organizations, that is). Not surprisingly, the need for this rarely arises, but we include it for completeness.

We saw that `owl:intersectionOf` corresponds to logical and, and `owl:unionOf` corresponds to logical or. The construct, `owl:complementOf` corresponds to logical negation. If an individual is in the set, it is not in the complement of the set, and vice versa.

Sanctioned Inferences

The main inferences that are sanctioned are as follows. For classes `C1` and `C2`:

```
IF:    C1 owl:complementOf C2.
THEN: C1 owl:disjointWith C2.
```

From the meaning of disjoint classes, we can further conclude that:

```
IF:    x rdf:type C2. and y rdf:type C1.
THEN: ¬(x rdf:type C1) and ¬(y rdf:type C2). and
             x owl:differentFrom y.
```

The latter triple introduces a new construct that we have not seen before. Its meaning is self evident, but the need to say it may not be. It follows from the fact that OWL does not make the unique name assumption. It is possible for the same individual to have two IRIs. If you know that is the case, you use `owl:sameAs` to link them; we saw this in Figure 4.13. Note that `owl:sameAs` and `owl:differentFrom` are closely related.

```
IF:    x owl:sameAs y.
THEN: ¬(x owl:differentFrom y) and vice versa.
```

Exercise 1: Create Venn diagrams to convince yourself that the inferences for intersection, union, and complement are justified.

5.3.3   ENUMERATION

In all of our examples so far, we have talked about creating classes and then declaring that or determining whether a given individual is or is not a member of that class. We never spoke of a limit as to how many members a class may have.

Sometimes you do know exactly what the members of a class are beforehand, e.g., when placing things into categories. In the U.S., there is a distinction between exempt and non-exempt employees. So you might have a class called `ExemptType` with just two members corresponding to those two categories.

Another example is the set of the base units in the International System of Units. These include meter, second, candela, ampere, and a few others. There aren't that many, and they are not adding new ones any time soon.

The OWL construct for this is called `owl:oneOf`. It works like this (in Turtle syntax):

```
doe:ExemptType rdf:type owl:Class .
doe:ExemptType owl:equivalentClass
                [ rdf:type owl:Class ;
                owl:oneOf(doe:_nonExempt
                        doe:_exempt) ].
```

Recall that anonymous classes are expressed in Turtle using square brackets and lists are enclosed in round parentheses. The expression using `owl:oneOf` returns a class with the two members specified. Note that the structure of the Turtle is just like that for intersection and union, except that `oneOf` is being used in place of `intersectionOf` or `unionOf`. Instead of having a list of classes that are all in an intersection or union, there is a list of individuals in the enumerated class.

The Manchester syntax for enumeration is to enclose the individuals in squiggly brackets and separate with commas, like this: `{doe:_exempt, doe:_nonExempt}`.

Sanctioned Inferences

An enumerated class is closed. This means that if an individual is known to be different from all the enumerated members of a class, then it cannot be a member of the class This is another way of saying it is in the complement of the enumerated class.

IF: `C2 owl:equivalentClass {x1,x2, …,xn}.` and
    `C1 owl:complementOf C2.` and
      for i=1,n `x owl:differentFrom xi.`
THEN: `x rdf:type C1.`

Also, if any resource, `y`, is in fact referring to the same thing as one of the members of the enumerated class, then it is inferred to be a member of the enumerated class.

IF: for any i, `y owl:sameAs xi`
THEN: `y rdf:type C2.`

5.3.4 PROPERTY RESTRICTIONS

Another kind of OWL class expression is used to create property restriction classes, which define a class based on how many individuals (or literals) it is connected to by a given object (or data) property. We saw a few examples in Section 2.3.4 (e.g., `ThingWithTwoWheels`). Restrictions are covered in the next major section.

5.3.5 SUMMARY: CLASS EXPRESSIONS

1. Classes may be defined as expressions using a variety of operators.

2. The main ones are existential and cardinality property restrictions (described in the next section) and the Boolean operators union and intersection.

3. A class expression consists of various combinations of classes, properties, and individuals and the result is a new class which may be used in any OWL expression anywhere a class is expected. For example:

   a. as a domain or a range,

   b. as a filter class in a property restriction, and

   c. in any other Boolean expression.

4. It is possible to nest expressions to arbitrary levels, but doing so gets confusing quickly. Keep things simple.

5. A class expression is not itself a triple, but it can be used in triples where a class is expected. Expressions show up in triples as blank nodes, as depicted in Figures 5.1 and 5.2.

5.4 PROPERTY RESTRICTIONS

The Boolean expressions in the prior section are used to formally represent something about what it means to be a member of the class, as reflected by specific inferences that are sanctioned. To be a member of the class `Motorcycle`, an individual must be a member of two different classes: `TwoWheeledVehicle` and `MotorizedVehicle`. This is important, but it begs the questions: what does it mean to be members of those classes? Remember, as a human, you can read into the meaning of the names, but the computer cannot. We have to spell it out.

Central to understanding the meaning of a class is to know what properties members of that class have. For example, every two-wheeled vehicle has two different wheels as parts. To be a patient, an individual must have received care on some patient visit.

In this section we explore a variety of ways that OWL provides to give meaning to classes based on the nature and number of properties that members of the class may, must or must not have. Classes defined this way are called property restrictions. This idea was introduced with several examples in Section 2.3.4.

Below are examples of classes that are naturally defined as restrictions. They exemplify the six main types of restrictions provided by OWL. For each we give the Manchester syntax of one way to model the restriction; it is just what you could type into an ontology editor to create these restrictions.

Existential: The set of all things that cover some party (in the context of insurance).
```
(doe:cover some doe:Party)
```

Universal: The set of all things that have only physicians as members.
```
(doe:hasMember only doe:Physician)
```

Minimum cardinality: The set of all things that are verified by at least 2 people.
```
(doe:verifiedBy min 2 doe:Person)
```

Maximum cardinality: The set of all things that have no more than 3 registered devices.
```
(doe:registeredOn max 3 doe:Device)
```

Exact cardinality: The set of all things that have exactly two wheels.
```
(doe:hasPart exactly 2 doe:Wheel)
```

Individual value: The set of all things that are part of Sentara Healthcare.
```
(doe:partOf value doe:_SentaraHealthcare)
```

There are other variations that we will describe later in this section.

5.4.1   USAGE SCENARIOS

Because a restriction is a class, it can be used anywhere that classes are used. That includes domain and range as well as anywhere in an expression that uses any combination of intersection, union or complement. However, by far the most common way to use a restriction is when you have a class in mind and you want to add meaning to it using either `equivalentClass` or `subClassOf`. If you have a restriction in mind and want to name it with an IRI, you use class equivalence and the restriction provides an exact meaning of the class. If instead you are just interested in ensuring that every member of the class has a particular property, you use subclass.

Restrictions Using Class Equivalence

Class equivalence is deployed when you want to use a restriction as an expression, or as part of an expression that you wish to give a name (i.e., an IRI). A good reason to do this is if you expect to use the class in a number of other places, e.g., in domains and ranges or in larger class expressions.

Earlier we used the class `TwoWheeledVehicle` in an intersection to define the class `MotorCycle`. Using a restriction, we can be more explicit about what that means. Delving deeper, a two-wheeled vehicle is both a vehicle and something with two wheels. The latter is naturally represented as an OWL property restriction.

In English, the class `TwoWheeledThing` is the set of all things that have exactly two parts that are wheels. If there is certainty about having more or fewer than two wheels, an individual is ruled out of the class. In terms of triples, the class `TwoWheeledThing` is the set of individuals that are the subjects of exactly two triples using the `hasPart` predicate where the objects of the triples are different from each other and both are of type `Wheel`. The differentness criteria is important. Two copies of the exact same triple do not count as two wheels. Also, because there is no unique name assumption, if there were two different IRIs for a given wheel then there could be two triples pointing to the same wheel (see Figure 5.3). What is important but is not shown in the figures, is that the pairs (w1, w2), (w3, w4), and (w5, w6) each refer to distinct wheels.

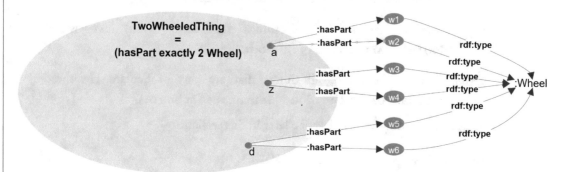

Figure 5.3: Property restrictions using class equivalence.

In principle, the usage pattern is:

1. create a class;

2. create a property restriction; and

3. make the class equivalent to the property restriction.

In practice, ontology-authoring tools often combine steps 2 and 3. For example, in Protégé, you select the class, you click to indicate you wish to create an equivalent class, and then you specify the restriction as an expression (see Figure 5.4).

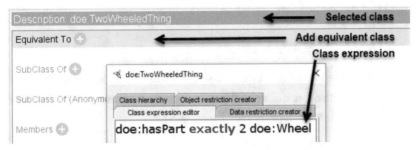

Figure 5.4: Specifying class equivalence in Protégé.

The result in Manchester syntax is:

```
Class: doe:TwoWheeledThing
    EquivalentTo:
        doe:hasPart exactly 2 doe:Wheel
```

When using equivalence, the inference goes both ways. Specifically:

1. if you know an individual is a member of the class, `TwoWheeledThing`, you know it is connected to exactly two (different) wheels via `hasPart`; and

2. conversely, at least in principle, if you know that an individual is connected to exactly two (different) wheels via `hasPart` then you know it is a member of the class `TwoWheeledThing`. In practice, due to the open world, it is generally not possible to be sure. There might be other wheels not known about.

Exercise 2: Spell out the inferences regarding `TwoWheeledThing` in a more precise notation using triples.

Restrictions Using Subclass

The second common way to use a restriction is very similar to the first way, except we use subclass instead of equivalence. This achieves the goal of saying that every member of a particular class *necessarily* has a particular property. It does not say that having that property is sufficient to warrant being a member of that class. In other words unlike equivalence, the inference only goes one direction. For example, an insurance policy necessarily covers one or more parties, be they persons or organizations. To proceed, we create a property restriction using the expression as: "`covers some Party`," which refers to "the set of all things that cover some party." Then we make `InsurancePolicy` a subclass of this property restriction class. In Manchester syntax:

```
Class: doe:InsurancePolicy
    SubClassOf:
        doe:covers some doe:Party
```

Because of the meaning of subclass, if an individual is a member of the class `Insuran-cePolicy`, then they are also a member of the class "`covers some Party`," which means every insurance policy covers at least one party (see Figure 5.5).

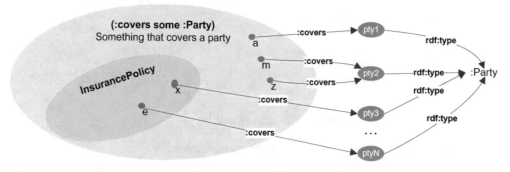

Figure 5.5: Property restrictions using subclass.

The inferences sanctioned by this usage scenario are:

IF: `x rdf:type :InsurancePolicy`.
THEN: there exists a `y` such that

     `x :covers y`. and
     `y rdf:type :Party`.

The usage pattern is:

1. create a class;

2. create a property restriction; and

3. make the class a subclass of the property restriction.

This is accomplished in Protégé just like the equivalent usage scenario, the only difference is that you click the plus sign next to where it says `SubClassOf` rather than where it says `Equiv-alent To` (see Figure 5.6).

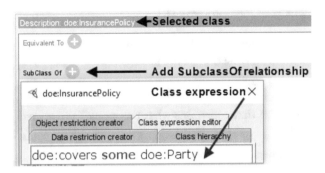

Figure 5.6: Specifying subclass relationship in Protégé.

Summary: Property Restriction Usage Scenarios

There are two main usage scenarios for property restrictions. Both entail having a class in mind that you want to add meaning to, and in each case you create the class and you create the restriction. If you are just interested in ensuring that every member of the class has a particular property, you use subclass. If instead you want to give the restriction a name and have the flexibility to use it in other expressions, you use class equivalence. A good example of this is `Party`, which we defined to be equivalent to the union of `Person` and `Organization`. You can just use the class `Party` as a domain or range or filter class on a restriction instead of repeating the expression "`Person or Organization`" over and over.

   This is very important: read those few sentences again and review the material in this section and at the beginning of Section 2.3.4, in particular, Figure 2.8 until you understand it.

   In addition to these two main usage scenarios, keep in mind that because a restriction is a class expression, it may be used anywhere that classes are expected. We already saw one example of this in Figure 2.10 where we defined `doe:Patient` to be equivalent to the expression:

`doe:Person and (doe:careRecipientOn some doe:PatientVisit).`

   Here the restriction is being used as one of the arguments in a larger intersection expression. You can also have a restriction be the filter class in another restriction. We will see an example of this in Chapter 6.

5.4.2   ANATOMY OF A PROPERTY RESTRICTION

A restriction denotes a set of subjects of triples using a given predicate that satisfy certain conditions. I think of these triples as the restriction's grounding triples. They are depicted in thick lines in Figure 5.7. The subjects of the grounding triples in the figure are s1, s2, … sN. Each property restriction has three parts.

1. *Property:* An object or data property IRI.

2. *Type:* The type of restriction. We just saw two examples, each indicated by a specific keyword. Existential restrictions use the keyword "`some`" and exact cardinality restrictions use "`exactly N`."

3. *Filter:* A qualification that constrains what the type of the objects of the restriction's grounding triples may or must be. The filter is optional for some of the restrictions, and required for others.

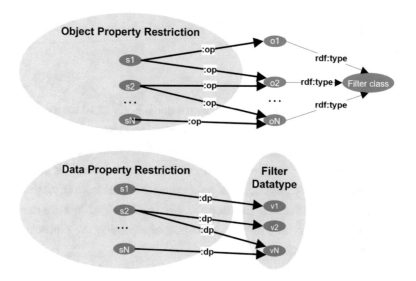

Figure 5.7: A restriction is defined in terms of triples.

The first part is straight-forward. The main point of interest is whether it is an object property or a data property. There are six main types of restriction, as laid out in the beginning of this section: existential, universal, three cardinality restrictions (minimum, maximum, and exact), and individual value. The cardinality restrictions require a number to be specified. For example, the restriction that defined `TwoWheeledThing` was an exact cardinality restriction using the number 2.

For the first five restriction types, the filter is a class for object restrictions and a datatype for data restrictions. For individual value restrictions, the filter is either an individual or a literal, depending on whether the restriction uses an object or data property, respectively. For brevity, we will at times shorten the terms "data property restriction" and "object property restriction" to "data restriction" and "object restriction."

Qualified Cardinality

The filter is required for universal, existential and individual value restrictions. It is optional for the cardinality restrictions. If a filter is not specified for an object property restriction, it is exactly the same as if the filter class was `owl:Thing`. When the filter is specified for a cardinality restriction it is called a *qualified cardinality restriction*. When using common ontology editors to build ontologies, you may not see the term "qualified cardinality" anywhere. You also will not see it if you export the ontology in Manchester syntax. However, it is there if you export it as Turtle or RDF/XML, or if you load it into a triple store.

The anatomy of a restriction as described above is reflected in the Turtle syntax for our two examples below as well as in Figure 5.8.

```
doe:TwoWheeledVehicle
    rdf:type owl:Class ;
    owl:equivalentClass
        [ rdf:type owl:Restriction ;
          owl:onProperty doe:hasPart ;
          owl:onClass doe:Wheel ;
          owl:qualifiedCardinality
                  "2"^^xsd:nonNegativeInteger
        ] .


doe:InsurancePolicy
    rdf:type owl:Class ;
    rdfs:subClassOf [ rdf:type owl:Restriction ;
                      owl:onProperty doe:covers ;
                      owl:someValuesFrom doe:Party
                    ] .
```

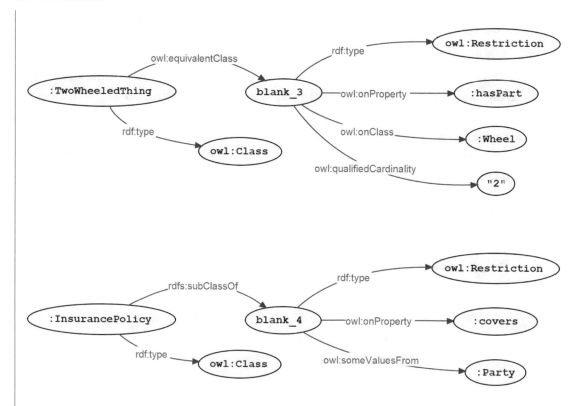

Figure 5.8: Restrictions as triples using blank nodes.

The class expression in Manchester syntax, "`covers some Party`," shows up in Turtle syntax as an expression inside square brackets that represents the anonymous class as a blank node. This anonymous class is an instance of `owl:Restriction`. The construct `owl:onProperty` is used to specify the property IRI. The construct `owl:someValuesFrom` is used to specify the type of restriction (existential), as well as the filter class `Party`. Manchester syntax is more human readable, Turtle is closer to the actual triples, which is important if you plan to be using SPARQL to query a triple store using the ontology.

Notice that the keyword `owl:qualifiedCardinality` is used in the first example. There are two aspects to this. First, it needs to be qualified cardinality because there is a filter class (in this case, `Wheel`). Second, this construct means the number is exact, rather than a minimum or maximum. The OWL keywords for exact, min and max cardinality are `owl:qualifiedCardinality`, `owl:minQualifiedCardinality`, and `owl:maxQualifiedCardinality`, respectively. A more informative keyword for the first construct would have been: `owl:exactQualifiedCardinality`. We will look into each of these in the sections below.

Notice also that although the filter is in the same place for all restriction types in Manchester syntax, it is different for Turtle. The predicates indicating the type of restriction for existential, universal, and individual value restrictions are `owl:someValuesFrom`, `owl:allValuesFrom`, and `owl:hasValue`. The objects of the triples using these three predicates to define restrictions are all filters. Due to the names of the constructs, people frequently use the terms "some values," "all values," and "has value" instead of "existential," "universal," and "individual value." We will follow that convention. For the cardinality restrictions, the predicates indicating the type of restriction are: `owl:qualifiedCardinality`, `owl:minQualifiedCardinality`, and `owl:maxQualifiedCardinality`. The objects of triples using these predicates to define restrictions are non-negative integers.

Object vs. Data Property Restrictions

Data properties are used in the same way as object properties when creating restrictions. The main difference is that the filter is a datatype, not a class. For the cardinality restrictions, the filter is specified separately, using `owl:onClass` for object property restrictions and using `owl:hasDataRange` for data property restrictions.

Figure 5.7 accurately reflects all but one of the restriction types. Not covered are individual value restrictions; they are described in Section 5.4.8. Note that a member of a restriction class can sometimes have more than one triple using the property that matches the filter. That is a question of cardinality.

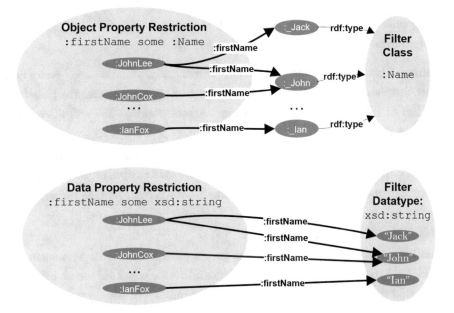

Figure 5.9: Examples of object and data restrictions.

In every case the restriction defines a class whose members may, must, or must not be the subject of certain numbers of triples using the property as the predicate and the object obeying the filter. In the next sections we will look at these six kinds of OWL restrictions. Each kind works for both object and data properties.

Figure 5.9 depicts two ways to represent the set of individuals with first names. They are examples of the general patterns shown in Figure 5.7. One uses a data property, the other uses an object property. The thicker lines are the grounding triples for the respective restrictions.

5.4.3 EXISTENTIAL: SOMEVALUESFROM

We have seen several examples of this type of restriction already, including the insurance policy example above. In Manchester syntax, the form is: `(p some C)` where `p` is a property and `C` is a class.

Sanctioned Inferences

IF: `x rdf:type (p some C).`
THEN: there is some individual `y` such that
          `x p y.` and `y rdf:type C.`

In the insurance example, this corresponds to:

IF: `x rdf:type (:covers some :Party).`
THEN: there exists at least one individual `y` such that
          `x :covers y.` and `y rdf:type :Party.`

The phrase "there exists" sheds light on why this is called an "existential" restriction. To represent the fact that every insurance policy covers at least one party, we make the class for insurance policy a subclass of the restriction class, "`covers some Party.`" We will discuss this kind of inference more in Section 6.4.2.

For data restrictions, it works the same way. The difference is that the filter is not a class, `C`, but a datatype, `D`. Also, the typing of a literal does not use `rdf:type`, rather it is appended to the string connected by the characters "^^", e.g., `"John"^^xsd:string` (see Figure 5.9).

5.4.4 UNIVERSAL: allValuesFrom

In healthcare, physicians often band together as groups to negotiate better agreements with insurance companies. So there is a class called `PhysicianGroup`. We could use an existential restriction to ensure that there is at least one physician as a member. We also don't want anything *else* to be a member of a physician group. This is what the universal restriction is for. We want to say that a physician group *only* has physicians as members. In Manchester syntax, the form is: `(p only C)` where `p` is a property and `C` is a class.

Sanctioned Inferences

IF: `x rdf:type (p only C).`
THEN: For all individuals `y`
        IF: `x p y.`
        THEN: `y rdf:type C.`

In English, it means that a member of the restriction class `(p only C)` can be connected via property `p` *only* to individuals of type `C`. In the physician example, this expands to:

IF: `x rdf:type (:hasMember only :Physician).`
THEN: For all individuals `y`
        IF: `x :hasMember y.`
        THEN: `y rdf:type :Physician.`

By making `PhysicianGroup` a subclass of this restriction, we achieve the intended effect. The use of the term "for all" sheds light on why this is called the universal restriction.

Note that due to the open world, you cannot infer an individual to be a member of an all values from restriction. You could have hundreds of members of a group and all of them might be physicians, but the inference engine leaves *open* the possibility that there are non-physician members that it does not know about.

Must a Physician Group Have Members?

A question arises in a situation like this as to whether you want to allow an individual resource to be a `PhysicianGroup` if it has no members at all. Suppose you don't. It turns out that the above inference rule for universal restrictions says nothing about this case, it only says what must be true if there *are* members.

So using a universal restriction on its own, does allow a `PhysicianGroup` with no members. To force at least one member, just add an existential restriction. You end up making the class `PhysicianGroup` a subclass of each of the following restrictions "`hasMember some Physician`" and "`hasMember only Physician`." This is a commonly used pattern.

5.4.5   MINIMUM CARDINALITY

Minimum cardinality is a generalization of the `someValuesFrom` restriction. To say that there is some `hasMember` triple is to say that there is a minimum of 1 `hasMember` triples. Rather than just being able to say there is a minimum of one, you can use any non-negative integer. In Manchester syntax, the form is: (p min n C) where p is a property and C is a class.

Sanctioned Inferences

IF: x rdf:type (p min n C).
THEN: there is a minimum of n different individuals y1 ... yn such that for i = 1, n
         x p yi. and yi rdf:type C.

Let's say that there is a requirement that a physician group has at least 20 members. We will make PhysicianGroup a subclass of the restriction (hasMember min 20 Physician). The general case expands as follows for this example:

IF: x rdf:type (:hasMember min 20 :Physician).
THEN: there is a minimum of 20 *different* individuals, y1 ... y20 such that for i = 1,20
         x :hasMember yi. and yi rdf:type :Physician.

Other Remarks

Note that the filter class is optional for all the cardinality restrictions. If you leave it off, it is the same as if you used owl:Thing for an object restriction. For a data restriction the default filter is rdfs:Literal. When you do use the filter class (or datatype), the restriction is said to be a "qualified cardinality restriction."

Note that although a minimum 1 restriction is logically identical to using a some values restriction, the inference engines treat them differently and there are some useful things you can do with someValues restrictions that you cannot do with min cardinality restrictions. Thus, it is better to use some values rather than min cardinality. This is explained more fully in Section 7.6.

5.4.6   MAXIMUM CARDINALITY

Sometimes it is useful to limit the number of relationships of a certain type that an individual may have. For example, software licenses often limit use to a specified number of devices. Suppose there were three tiers of licenses for a large software firm. The basic tier allows just 1 device, the small business tier allows up to 5, and the enterprise tier allows an unlimited number. One way to model the middle tier is to create a subclass of SoftwareLicense called SmallBusinessLicense and make it also be a subclass of a restriction that means "the set of things that are registered to a maximum of 5 devices" (see Figure 5.10). Notice that the Turtle syntax below explicitly calls out the fact that this is a qualified cardinality restriction.

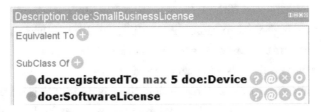

Figure 5.10: Max qualified cardinality restriction in Protégé.

```
doe:registeredTo rdf:type owl:ObjectProperty .

doe:Device rdf:type owl:Class .

doe:SmallBusinessLicense

   rdf:type owl:Class ;

   rdfs:subClassOf doe:SoftwareLicense ;

   rdfs:subClassOf

       [ rdf:type owl:Restriction ;

         owl:onProperty doe:registeredTo ;

         owl:onClass doe:Device ;

         owl:maxQualifiedCardinality

              "5"^^xsd:nonNegativeInteger

       ] .
```

In Manchester syntax, the form is: (p max n C) where p is a property and C is a class.

Sanctioned Inferences

IF: x rdf:type (p max n C).
THEN: there can be no more than n different individuals y1 ... yn such that for
        i = 1,n
        x P yi. and yi rdf:type C.

When applied to the software example, we get the following.

IF: x rdf:type (:registeredTo max 5 doe:Device).
THEN: there can be no more than 5 different individuals y1 ... y5 such that for
        i = 1,5
        x :registeredTo yi. and yi rdf:type :Device.

In practice, this means that if the inference engine can independently verify that that a particular license is already registered to more than 5 different devices then the license cannot be in the class, `SmallBusinessLicense`. If someone comes along and asserts that license into class `SmallBusinessLicense` there is a logical contradiction.

Note that unlike a min cardinality restriction, there is no way to infer that an individual is a member of a max cardinality restriction. This is due to the open world. The inference engine might only know about 5 device registrations, but there could be more out there.

Max 1 Cardinality vs. Functional Property

Max cardinality will be mostly used with n=1. This is very similar to functional properties that also have a maximum of 1 value. It turns out that declaring a property p to be functional is precisely to say that `owl:Thing` is a subclass of the property restriction (`p max 1`). Specifically, the following two assertions are identical in meaning:

1. `:hasOfficialName rdf:type owl:FunctionalProperty`

2. `owl:Thing rdfs:subClassOf (:hasOfficialName max 1)`

Exercise 3: Can you convince yourself that asserting a given property, p, to be functional means the same thing as asserting that `owl:Thing` is a subclass of the restriction (`p max 1`)?

5.4.7   EXACT CARDINALITY

Sometimes you know beforehand exactly how many relationships of a given kind a member of a given class always has. Two examples we have previously seen are:

1. a patient visit has exactly one person receiving care; and

2. a bicycle has exactly two wheels.

Exact cardinality is a shorthand for the combination of minimum cardinality and maximum cardinality. Specifically, the following two class expressions are equivalent:

1. `(hasPart min 2 Wheel) and`
   `(hasPart max 2 Wheel)`

2. `(hasPart exactly 2 Wheel)`

Because it is just a short hand, the inferences sanctioned are just those that are described above. We do not repeat them. However, I do repeat the fact that the open world prevents inferring an individual to be a member of a max cardinality restriction. And because an exact cardinality restriction combines a min *and* a max, the inference engine will also never be able to infer an individual into an exact cardinality restriction.

5.4.8 INDIVIDUAL VALUE: `hasValue`

It is common in an enterprise to refer to internal organizations vs. external organizations. What does that mean? Essentially, it means that the organization is or is not ultimately a part of the enterprise itself. For example, `_SentaraHealthcare` is a `Corporation`. A `SentaraOrganization` is an organization that ultimately is a part of `_SentaraHealthcare`. This could include informal departments, right up to major subsidiaries. We want to create a property restriction that means: "something that is a part of Sentara Healthcare, a particular company." This is very similar to a `someValuesFrom` restriction, but the filter is not a class, it is a particular individual.

It works for data properties in exactly the same way, the only difference is the filter is a literal, not an individual. As an example, we could use the subclass pattern in conjunction with `hasValue` to represent that all members of the class `RedCar` have the color red, where colors are represented as strings (see Figure 5.11 and compare it to Figure 5.9). Note that the keyword for an individual value restriction in Manchester syntax is value. For example: `:hasColor value "red"^^xsd:string`, as shown in Figure 5.11.

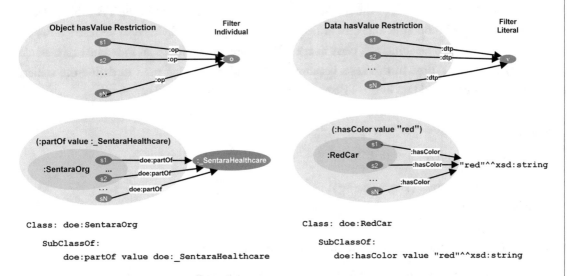

Figure 5.11: Individual value restrictions with grounding triples.

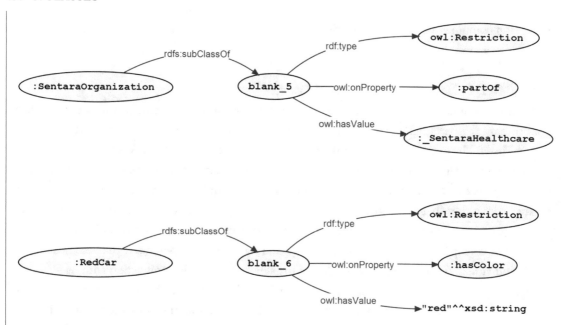

Figure 5.12: Defining and using individual value restrictions.

Note that until now all of the restriction types have been specified in the TBox only, that is by referring only to properties and classes from the ontology. The individual value restrictions are unique in that to give meaning to a class in the TBox, they must refer to an individual that would normally be in the ABox. In this case, because the individual is being used to define the subject matter, it probably should go in the TBox with other classes and properties.

Sanctioned Inferences

For individuals x and v:

IF: x rdf:type (p value v).
THEN: x p v

In the internal Sentara organization example, this corresponds to:

IF: x rdf:type (doe:partOf value :_SentaraHealthcare).
THEN: x doe:partOf :_SentaraHealthcare.

5.4.9   DATA PROPERTY RESTRICTIONS

Most of the above examples and illustrations use object properties. However, all of the six kinds of restriction can be used with either object properties or data properties. It works exactly the same way and the inferences are exactly analogous. The main difference is in the filters. Except for the individual value restrictions, the filter for a data property restriction is a datatype rather than a class. For the individual value restrictions, the filter is a literal value rather than an individual. Below we list the general form of a data restrictions for each of the six kinds of restriction, including an example. The symbols D and v below denote datatype and literal value, respectively.

1. Some values: `(p some D)`
   e.g., `(:firstName some xsd:string)`

2. All values: `(p all D)`
   e.g., `(:hasSSN all xsd:integer)`

3. Min cardinality: `(p min n D)`
   e.g., `(:hasDescription min 1 xsd:string)`

4. Max cardinality `(p max n D)`
   e.g., `(:hasLicenseNumber max 1 xsd:string)`

5. Exact cardinality `(p exactly n D)`
   e.g., `(:hasCodeName exactly 2 xsd:string)`

6. Has value: `(p value v)`
   e.g., `(:hasColor "red"^^xsd:string)`

Exercise 4: Write down in English the meaning of each of the above data property restrictions.

5.4.10 SUMMARY: PROPERTY RESTRICTIONS

What Is a Property Restriction?

A *property restriction* is a class that is defined in terms of what relationships individuals of that class may or must have with other individuals (for object property restrictions) or with literals (for data property restrictions). Having a relationship means being the subject of a triple.

A *filter* can be specified to indicate what type of individual or literal the object of the triple must be. It is specified as a class for an object restriction, and as a datatype for a data restriction. Such filters are required for all values and some values restrictions. They are optional for cardinality restrictions. A cardinality restriction that uses a filter is called a qualified cardinality restriction.

Using Restrictions

There are two main scenarios for using restrictions to capture meaning of the classes in your ontology. The first is via class equivalence. In this scenario, you are effectively giving a name to a property restriction which by itself is an anonymous class represented as a blank node. More specifically, you are creating a separate IRI that can be used instead of the blank node. This sanctions the following two inferences. First, if you know an individual is a member of the class, then you can infer that it has properties of the sort dictated by the restriction. Second, if you have an individual that meets the conditions of the restriction, you can infer it into the class.

The second scenario for using property restrictions is via the subclass relationship. You create the restriction and make your class a subclass of the restriction. This achieves the goal of saying that every member of a particular class *necessarily* has a particular property without saying that everything that has that property is a member of that class. In this case, inference only goes one way.

In addition to the above usage scenarios for individual restrictions, because a restriction is a class, it may be used anywhere that a class is used in any other specification or expression.

5.5   SUMMARY LEARNING

Classes and Sets

A class represents a mathematical set of individuals that share certain properties. Venn diagrams are a helpful way to visualize class expressions and relationships.

Class Expressions

Classes may be defined as expressions using a variety of operations. The main ones are existential and cardinality property restriction operators and the Boolean operations, union, and intersection. A class expression combines classes, properties, and individuals and the result is a new class that may be used in any OWL expression where a class is expected. A class expression is not itself a triple, but it can be used in triples.

Class Relationships

Classes may be related to other classes in three main ways.

Subclass: The most common is the subclass relationship that specifies that the set of member individuals of one class is a subset of the set of member individuals of the other class. This relationship gives rise to a class hierarchy.

Class equivalence: Two classes are equivalent if both sets have exactly the same member individuals. The most common use of class equivalence is to specify additional meaning using a class expression. Another use is for mapping two ontologies.

Disjoint classes: Two classes being disjoint means that there are no individuals that are members of both classes. In addition to specifying more meaning for humans to understand the ontology, disjointness plays an important role in catching logical errors in an ontology.

Property Restrictions

Object property restrictions are used to state how many relationships of what kind an individual of a given class may have with other individuals that are members of another specified class.

Data property restrictions are used to state how many relationships of what kind an individual of a given class may have with literals of a given datatype.

A cardinality restriction indicates the number of relationships of a given kind that an individual may or must have with individuals or literals from a specified class or datatype.

A `someValues` restriction is equivalent to a min 1 cardinality restriction. It requires an individual to be in at least one other relationship of a specified kind with another individual in a specified class or datatype.

An all values restriction indicates that an individual can be related in a certain way only to individuals from a particular class or datatype.

Inference engines can infer individuals to be members of some values, min cardinality, and has value restrictions, but due to the open world, that is not possible for all values, max, or exactly restrictions.

5.6   CONCLUSION FOR PART 2

This concludes the meatiest part of the book. Congratulations, that was a lot of work! In the next and final part of the book we look into how to use OWL in practice.

Part 3

Using OWL in Practice

In Part 1, we explored what kinds of things need to be said and how to say them in OWL, in order to build an ontology. We also introduced the foundational concepts required to become a savvy user of OWL: sets, logic, inference, and meaning.

Part 2 took you on a thorough exploration of the variations and nuances of the most important OWL constructs related to properties and classes. The meaning of each construct was explained using a combination of English and more precise technical descriptions of what inferences are sanctioned by each construct.

While Parts 1 and 2 focused on *what* OWL is, the focus of Part 3 is on *how* to use it in practice. We consider a wider variety of practical situations that you will likely encounter when building OWL ontologies and the kinds of decisions you will be making day to day. We expand on prior examples and look at new ones. We point out some limitations of OWL and suggest workarounds. We conclude by giving some practical tips and guidelines and introducing some less frequently used OWL constructs that go beyond the 30% that you will be using 90% of the time.

CHAPTER 6

More Examples

The last two chapters went into a great amount of detail explaining the different ways to use OWL properties and classes to capture the semantics of important concepts in the subject matter of a given ontology. In this chapter we revisit some old examples with new insights, and we look at new examples showing how to put together all the knowledge about OWL we have discussed so far. We first present the examples, and what they look like as triples. After that we look at the kinds of inferences that you get with each of these examples.

6.1 PATIENT VISIT

Let's reconsider the patient visit example from Figure 2.9. Below we show the human-friendly Manchester syntax for OWL, and then the Turtle syntax which means exactly the same thing. Notice that the former is fairly easy to read, but the underlying triples are not evident. That's where Turtle shines. When you use your ontologies to develop applications, they will be loaded into a triple store and it will become more important to better understand things as triples.

In the Manchester syntax version, there is also no evidence of the four blank nodes that are used underneath the covers in a triple store. We introduced blank nodes in Section 5.3.1.

Manchester Syntax

```
Class: doe:PatientVisit
    EquivalentTo:
        doe:Event
          and (doe:careProvider some doe:Person)
          and (doe:careRecipient some doe:Person)
```

Turtle Syntax

```
doe:PatientVisit rdf:type owl:Class ;
  owl:equivalentClass
    [ rdf:type owl:Class ;
      owl:intersectionOf
        ( doe:Event
```

```
        [ rdf:type owl:Restriction ;

          owl:onProperty doe:careProvider ;

          owl:someValuesFrom doe:Person ]

        [ rdf:type owl:Restriction ;

          owl:onProperty doe:careRecipient ;

          owl:someValuesFrom doe:Person ]

    )

]
```

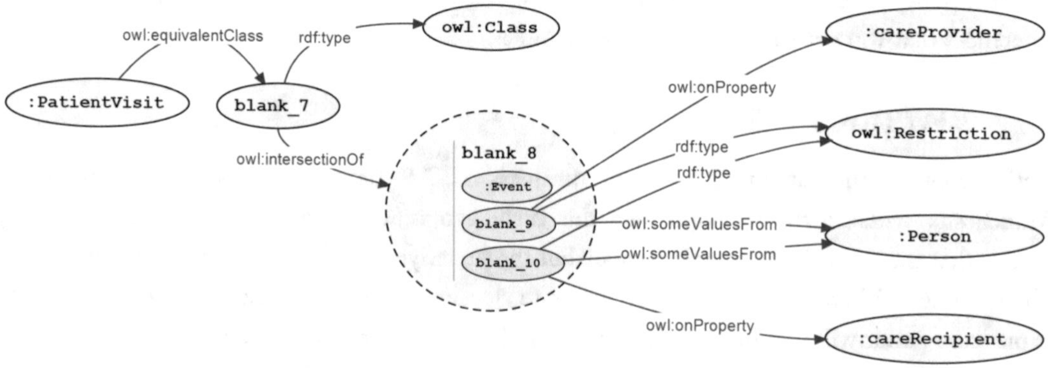

Figure 6.1: Patient visit as triples.

Figure 6.1 shows how all this unpacks as triples. We create the :PatientVisit class and set it to be equivalent to another class. That other class is represented as a blank node which in Turtle is enclosed by square brackets. When you encounter a left square bracket (i.e., "[") you can read it more or less as *"an anonymous class that is the subject of the following triples"* where the closing square bracket means there are no more triples for that blank node. For example, :PatientVisit is equivalent to *an anonymous class (called :blank_7) that is the subject of the two triples*:

```
:blank_7 rdf:type owl:Class .

:blank_7 owl:intersectionOf :blank_8 .
```

The blank node, :blank_8 is the list of classes that are to be intersected. The list is enclosed by round parentheses rather than square brackets. There is more detail that I am suppressing here about how lists are represented in RDF. It's pretty ugly and you shouldn't need to worry about it for a while.

Of the intersected classes, one is :Event and the other two are restrictions, each of which are also represented as blank nodes (numbered 9 and 10 in Figure 6.1).

Using the guideline from above on how to read a blank node in Turtle, let's try to read the first restriction in the list:

```
[ rdf:type owl:Restriction ;
  owl:onProperty doe:careProvider ;
  owl:someValuesFrom doe:Person ]
```

We read this as: "*an anonymous class (called :blank_9) that is the subject of the following three triples*:

```
:blank_9 rdf:type owl:Restriction .
:blank_9 owl:onProperty doe:careProvider .
:blank_9 owl:someValuesFrom doe:Person .
```

The names I am giving to the blank nodes are just for illustrative purposes; they need to be created and given distinct names internally, but they are temporary and local. They cannot be used anywhere else, unlike IRIs for non-blank nodes in an RDF graph.

6.2   COLLATERAL

A slightly more complicated example comes from the finance industry, specifically how to represent the fact that many loans are protected by a security agreement in the event that the borrower stops making payments. We will introduce the idea of a security agreement that is a specialization of a more generic written contract. We define a security agreement to be equivalent to a written contract that has associated collateral.

For car and home mortgages, the collateral is usually the car or the home, but it can in principle be any owned thing of sufficient value. Let's start by just saying there is collateral. Figure 6.2 shows one way to do this. In the figure, the blue text corresponds to the blue shape.

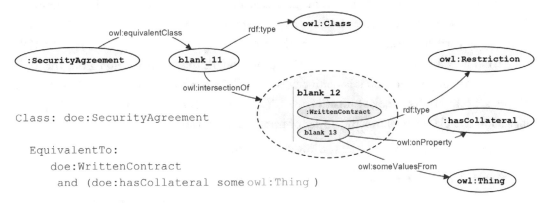

Figure 6.2: Security agreement.

We have said nothing about what the collateral can be, so the default is `owl:Thing` (in blue in the figure). One way to be more specific is to create a class called `Collateral`, but then you have a restriction that says "`hasCollateral some Collateral`." Setting aside that unsatisfying air of redundancy, collateral is not really a kind of thing. Something of value that is owned becomes collateral when used to limit the liability of a lender.

You want to focus on the meaning: what is true and important about the thing that is being used as collateral? It has to have some estimated value and it has to be owned. We can use property restrictions to say this without creating a class for collateral. We will extend the definition above by replacing the filter class on the `hasCollateral` restriction (`owl:Thing`) with a new class expression. We will create an OWL definition of `:SecurityAgreement` that means: "a written contract that is backed by an something with value that is owned." The filter class is an expression which is the intersection of the following two restrictions in Manchester syntax (see Figure 6.3).

1. `:hasEstimatedValue some :MonetaryAmount`

2. `:isOwnedBy some :IndependentParty`

This means that if an individual is asserted into the class, `SecurityAgreement`, then:

1. it is known to be an instance of `WrittenAgreement`; and

2. it is known to have collateral that is both owned and has an estimated dollar value.

Suppose we also require that a security agreement is always attached to a loan contract. We can add another restriction, but in this case we use subclass instead of equivalence. This extra restriction is not in Figure 6.3, but what it looks like in Protégé, e6Tools, and Manchester syntax is shown in Figures 6.4 and 6.5. Notice how Protégé uses Manchester syntax in its user interface. It mirrors the ontology when exported to a file (as shown in Figure 6.5).

Notice that the two intersected restrictions are part of the filter class for another restriction. The nesting of one or more restrictions inside another restriction sometimes called chaining restrictions. This works because a restriction is a class expression and a restriction can take any class as a filter. This particular example comes from the Financial Industry Business Ontology (FIBO).[23]

Even in Manchester syntax this takes a bit of thought to see what is going on. The Turtle would be more complicated, and it is not included here. Take some time to go over this a few times until you are comfortable. Look again at Figure 6.3. This example combines a number of important things all in one place and shows the power of using class expressions:

- both the subclass and the class equivalence patterns for using restrictions;

- nesting of an intersection within an intersection; and

---

[23] https://www.edmcouncil.org/financialbusiness.

• using an intersection of two restrictions as a filter nested within a restriction.

Note that the round parentheses in Manchester syntax are used purely for removing ambiguity, just like in arithmetic. They do not enclose a list, as they do in Turtle.

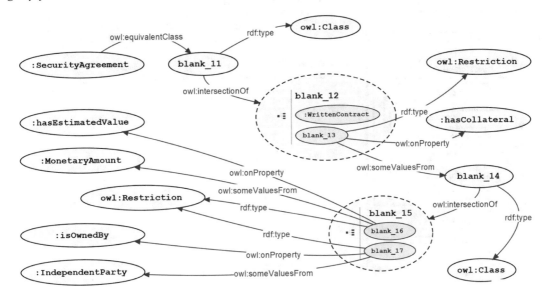

```
Class: doe:SecurityAgreement

EquivalentTo:
    doe:WrittenContract
    and (doe:hasCollateral some
          ((doe:hasEstimatedValue some doe:MonetaryAmount)
           and
           (doe:isOwnedBy some doe:IndependentParty))
        )
```

Figure 6.3: Security agreement, more details.

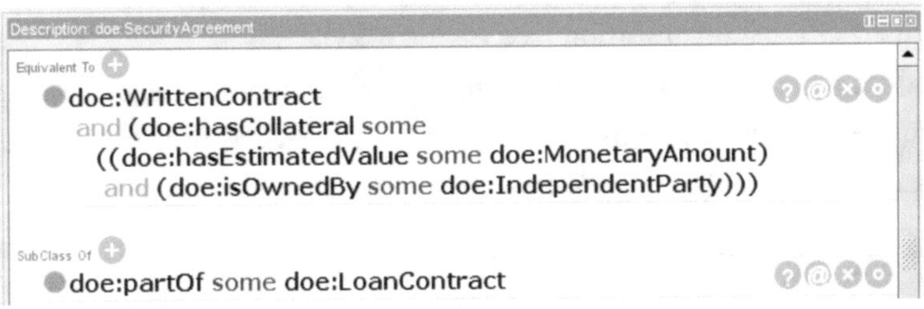

Figure 6.4: Security agreement; full details in Protégé.

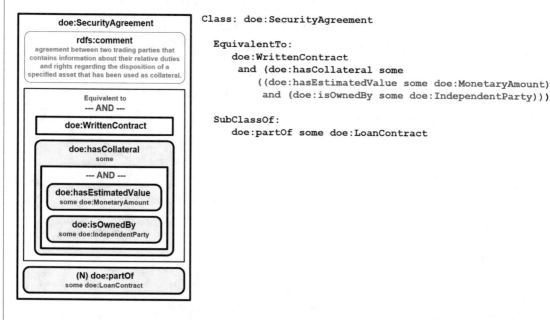

Figure 6.5: Security agreement in e6Tools and Manchester syntax.

6.3 INTERNAL VS. EXTERNAL TRANSACTIONS

Every enterprise engages in financial transactions with other companies or individuals regarding costs paid to vendors and revenue generated from sales. However, in any large company there are many internal organizations with separate budgets that transact business among themselves. For example, a company that provides cloud services to the public may also sell those same services to organizations within the enterprise. A single transaction is regarded as a cost by the purchasing organization and is regarded as revenue by the selling organization.

If you wanted to track internal financial transactions you might create a class defined to be a financial transaction where the buyer and seller were both internal organizations. We already talked about the idea of an internal organization (see Figure 5.11).

We introduce a company called Semvia that provides cloud services (among other things). We represent it as an instance of `:Organization` with IRI `:_Semvia`. We define the class, `:SemviaOrganization` to be equivalent to the intersection of two classes:

1. `:Organization`

2. `:partOf value :_Semvia`

The second class is a `hasValue` restriction. Note that `partOf` must be transitive to ensure that a lower organization in the company that has several layers of larger organizations above it is still regarded as being a part of Semvia. We will use class equivalence here because we want the inference to go both ways—that is, we want to infer into the class, as well as inferring information about something that is already known to be in the class.

To get to a definition of an internal Semvia transaction, we'll start by introducing a class for financial transaction that is an event where there is a seller, a buyer, and an amount of money that is paid. Also, the buyer and seller are both parties, where a party is defined to be either a person or an organization.

We create the class `FinancialTransaction` and make it a subclass of each of the following four classes:

1. `:Event`

2. `:hasBuyer some :Party`

3. `:hasSeller some :Party`

4. `:hasAmount some :MonetaryAmount`

We can now define a class for internal Semvia transactions where the buyer and seller are both internal Semvia organizations. It is defined as a financial transaction where the buyer and seller are both members of the class `SemviaOrganization` that we defined above. The class `InternalSemviaTransaction` is equivalent to the intersection of the following three classes:

1. `:FinancialTransaction`

2. `:hasBuyer some :SemviaOrganization`

3. `:hasSeller some :SemviaOrganization`

This structurally very similar to the definition of the patient visit class we just looked at. This is a common pattern that can be used for any subject.

Because we are using equivalence rather than subclass, the inference works both ways. First, if an individual is a member of all three classes, we can infer it to be an instance of the class `InternalSemviaTransaction`. Going the other way, if an individual is an instance of `InternalSemviaTransaction`, we can infer membership into each of the three classes.

The whole ontology that represents these ideas consists of eight classes and four object properties plus one inverse. See Figure 6.6 for a compact visual representation format. Most individuals, especially particular organizations would normally be in the ABox. However in this case, because: `_Semvia` is used to define a class, we include it in the TBox. Figure 6.7 shows how to populate this ontology with a few individuals. Each of the colorless rounded boxes represent individuals.

Figure 6.10 shows what this looks like in Protégé after running inference. See if you can explain to yourself exactly how and why the individual transaction, :_Trans_456 gets inferred into the class, :InternalSemviaTransaction. We will go through this in detail in Section 6.4.4.

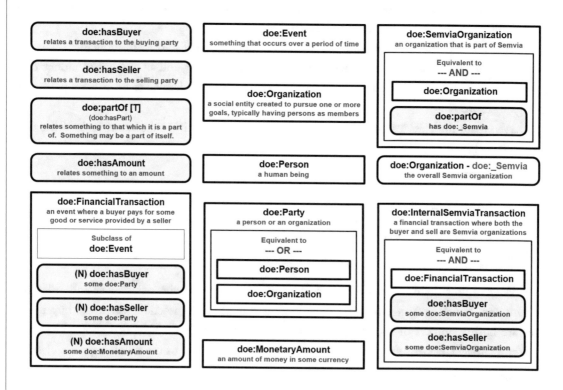

Figure 6.6: Internal organizations and transactions: TBox.

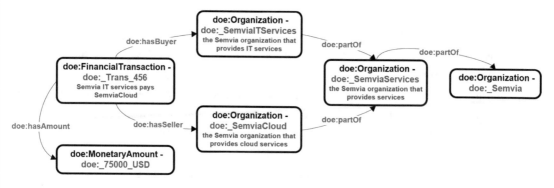

Figure 6.7: Internal organizations and transactions: ABox.

6.4   INFERENCE

Next, we consider what inferences can be drawn from the definitions in the previous examples.

6.4.1   PATIENT VISIT

Recall that with equivalence the inference goes both ways. If we assert an individual into the class, then we can conclude things about the individual from the definition. Conversely, if we assert certain things about an individual, we can conclude that it is a member of the class. The latter inference is depicted at the top of Figure 6.8. Every link corresponds to a triple, and the solid ones are asserted, the dotted ones are inferred. The heavy lines indicate what class an individual is an instance of.

The diagram tells us that if we assert that a given individual is an event, has a care provider, and has a care recipient (both of which are persons), then we can infer that individual into the class, `PatientVisit`. Conversely, if we assert the individual into the class, `PatientVisit`, then we can infer that: (1) the individual is a member of the class `Event`; (2) has a care provider that is a person; and (3) has a care recipient that is a person. See the bottom of Figure 6.8.

Because we don't know who the caregiver and care recipients are on the patient visit, the inference engine will temporarily make some distinct identifiers for those individuals that could potentially be used to draw other conclusions. This is called inference with partial information.

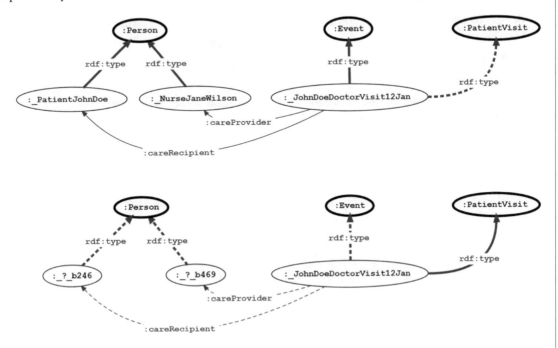

Figure 6.8: `PatientVisit` inference.

6.4.2 INFERENCE WITH PARTIAL INFORMATION

What is interesting about this is that we have no idea who those people are, we just know that they exist. This is analogous to knowing that Julius Caesar has a birthdate, even though you don't know what it is. In fact, no one knows. There is broad agreement among scholars that it was on July 12 or 13 and that it was one of the years 100, 101, or 102 BC.

We also saw this in the insurance policy example. We can infer the existence of a party that is covered even if we don't know exactly who that party is.

If we were modeling U.S. citizens we might wish to say that they all necessarily have a social security number. We could create a restriction and use the subclass pattern using a data property connecting people to their social security numbers as follows (in Manchester syntax).

```
Class: doe:Person

    SubClassOf:

        doe:hasSSN some xsd:integer
```

This enables us to infer that any individual known to be of type `:Person` will have a social security number, but we would generally not expect to store such information in a triple store.

6.4.3 SECURITY AGREEMENT AND COLLATERAL

At the top of Figure 6.9 we have asserted that a particular individual is a written contract, and also that it has collateral. Furthermore, the collateral has estimated value of $450k and is owned by Jane Will. This is sufficient to infer into the class `SecurityAgreement`. Convince yourself that this is true based on the above definition.

Once this happens, we can infer additional information based on the restriction (`partOf some LoanContract`) which security agreement is a subclass of. We know that it is part of a loan contract, even though we don't know which loan contract it is.

In the bottom part of the figure, we assert a single triple that places the individual `_secAgree_123` into the class `SecurityAgreement`. This triggers eight inferences, all depicted with dotted lines. The equivalence axiom in Figure 6.4 sanctions the six inferences on the left of the diagram. The subclass axiom sanctions the inference on the far right.

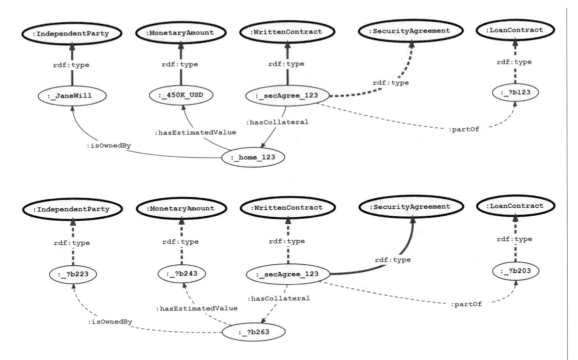

Figure 6.9: `SecurityAgreement` inference.

Exercise 1: Given our definition for a security agreement in Figure 6.5, what if any inference can be made if all you know is the following:

```
:_x doe:partOf :_LoanContract_203 .
:_?b203 rdf:type :LoanContract .
```

6.4.4 INTERNAL ORGANIZATIONS AND TRANSACTIONS

Figure 6.10 shows what the ontology in Figure 6.6 looks like in Protégé after running inference. Inferences show up with light yellow shading. You can get an explanation justifying the inference by clicking on the question mark to the far right. Figure 6.11 gives the details of why the individual, _Trans_456 is inferred into the class, `InternalSemviaTransaction`.

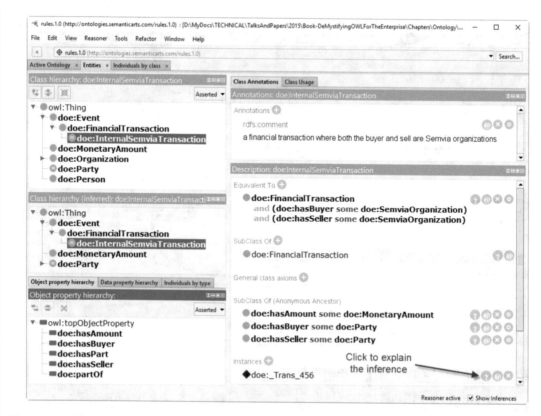

Figure 6.10: Internal organizations and transactions in Protégé.

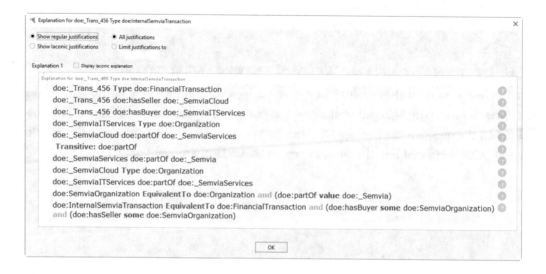

Figure 6.11: Explaining inference in Protégé.

We are going to unpack this. If we can determine that the individual `_Trans_456` is an instance of each of the three intersected classes making up the definition, then we can infer it to be in the class `InternalSemviaTransaction`. Those classes are depicted at the top of the lower right area of the Protégé window in Figure 6.10 and listed below.

1. `:FinancialTransaction`

2. `:hasBuyer some :SemviaOrganization`

3. `:hasSeller some :SemviaOrganization`

Look closely at the ABox depicted in Figure 6.7 to follow the reasoning. Each rounded shape is an individual, whose class IRI is in a bold black font. Next is the individual's IRI in blue. Below that in a smaller blue font is a comment. The same individuals in Turtle syntax are depicted in Figure 6.12.

First, we see that `_Trans_456` is directly asserted into the class `FinancialTransaction`. Next we need to show that `_Trans_456` is the subject of a triple with the predicate `hasBuyer` where the buyer is an instance of the class `SemviaOrganization`. The buyer is directly asserted to be `SemviaITServices`. Next we will try to show that `SemviaITServices` is a `SemviaOrganization`, which will enable us to conclude that `_Trans_456` is in fact a member of "`hasBuyer some SemviaOrganization`."

The definition of the class `SemviaOrganization` tells us that any instance that is an organization that is part of Semvia is a `SemviaOrganization`. So our specific strategy is to show that `_SemviaITServices` is a member of the following two classes:

1. `:Organization`

2. `:partOf value :_Semvia`

```
###################################################################
#    Individuals
###################################################################

doe:_75000_USD rdf:type doe:MonetaryAmount .

doe:_Semvia rdf:type doe:Organization ;
            rdfs:comment "the overall Semvia organization" .

doe:_SemviaCloud rdf:type doe:Organization ;
                 rdfs:comment "the Semvia organization that provides cloud services" ;
                 doe:partOf doe:_SemviaServices .

doe:_SemviaITServices rdf:type doe:Organization ;
                      rdfs:comment "the Semvia organization that provides IT services" ;
                      doe:partOf doe:_SemviaServices .

doe:_SemviaServices rdf:type doe:Organization ;
                    rdfs:comment "the Semvia organization that provides services" ;
                    doe:partOf doe:_Semvia .

doe:_Trans_456 rdf:type doe:FinancialTransaction ;
               rdfs:comment "Semvia IT services pays SemviaCloud " ;
               doe:hasAmount doe:_75000_USD ;
               doe:hasSeller doe:_SemviaCloud ;
               doe:hasBuyer doe:_SemviaITServices .
```

Figure 6.12: Internal organizations and transactions: ABox in Turtle.

_SemviaITServices is directly asserted to be an instance of Organization; it is also asserted to be a be a part of _SemviaServices, which is not quite what we want. We need it to be part of _Semvia. However, _SemviaServices is a part of _Semvia and by transitivity of partOf; we can conclude that _SemviaITServices is indeed a part of _Semvia. This allows us to infer it to be a member of the second class: "partOf value _Semvia." This sanctions inferring _SemviaITServices into the class _SemviaOrganization, which in turn allows us to infer that _Trans_456 is an instance of "hasBuyer some SemviaOrganization," the second of the three intersected classes defining InternalSemviaTransaction.

By the same line of reasoning, we can conclude that the seller _SemviaCloud is also a SemviaOrganization, which in turn allows us to infer that _Trans_456 is an instance of the third of the three intersected classes. We have now shown that _Trans_456 is a member of all three of the classes in its defining intersection. This is sufficient to infer it into the class InternalSemviaTransaction. Thus, we add the triple:

:_Trans_456 rdf:type :InternalSemviaTransaction.

The class SemviaOrganization gained the three new members: _SemviaCloud, _SemviaITServices, and _SemviaServices.

There is quite a lot going on here. It may take a few careful readings with a piece of paper before you follow all the details.

6.4.5 CLASSIFICATION INFERENCE

The many different inferences that are sanctioned by the various constructs combine to do some powerful things. The examples we have just seen involve automatically classifying individuals by inferring them into a class using the `rdf:type` construct. There are also many rules for inferring new `rdfs:subClassOf` links in the class hierarchy. The main ones are listed below and described in Section 5.3.2.

1. Inferring that a class defined as the intersection of two or more classes is a subclass of each of the intersected classes.

2. Inferring that a class defined as the union of two or more classes has each of the classes as subclasses.

A common pattern is the intersection of a named class and one or more property restriction classes. We defined `PatientVisit` to be the intersection of `Event` and two restrictions about giving and receiving care; therefore, by rule 1 above, `PatientVisit` is inferred to be a subclass of `Event`. Similarly, a `SecurityAgreement` is inferred to be a subclass of `WrittenContract`, and `Patient` is inferred to be a subclass of `Person`. Similarly, because `Party` is defined to be the union of `Person` and `Organization`, both of the latter classes are subclasses of `Party`.

The upper half of Figure 6.13 shows the top level classes before and after inference. `Organization` and `Person` are inferred to be subclasses of `Party`. The lower half of the figure shows both hierarchies fully expanded. Inference tidies things up a bit, putting things where they belong.

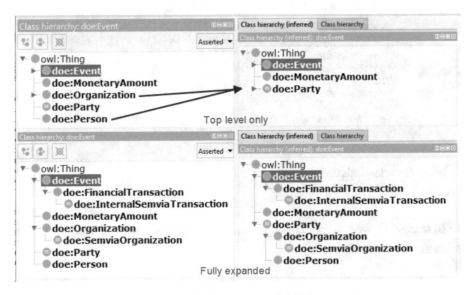

Figure 6.13: Class hierarchy inference.

Other more complex combinations also arise. For example, there is an ontology of generic enterprise concepts called gist.[24] In it, there is a class called `Intention` with four subclasses: `Goal`, `Permission`, `Requirement`, and `Restriction` (not an OWL restriction). In addition, the class, `Commitment` is defined to be the union of `Requirement` and `Restriction`. It turns out that `Commitment` is inferred to be a subclass of `Intention`. Any OWL DL compliant inference engine will draw that conclusion given the above information. Figure 6.14 depicts what happens to the hierarchy after running inference. The number of top-level categories is 31% smaller (from 26 to 18).

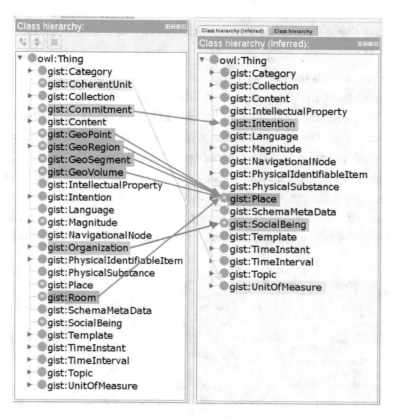

Figure 6.14: Classification inference in gist.

Question to reader: Draw a Venn diagram to make it easy to see why :Commitment is inferred to be a subclass of :Intention.

[24] www.semanticarts.com/gist. Disclosure: I have been using and co-developing gist since 2010.

Benefits of Classification Inference

There are two kinds of classification inference. One puts an individual into a class using `rdf:type`. The other results in adding new `rdfs:subClassOf` links between classes. This process is dynamic. When new triples are added, inference is run again. This updates the class hierarchy and may result in some individuals being members of new classes.

Classification inference helps in various ways. First, it saves work in maintaining the subclass hierarchy. Many relationships are added for you. If you make changes that have ripple effects, the inference engine takes care of that for you. This makes it easier to maintain the ontology. Another important benefit is that it helps to spot mistakes. For example, if `Commitment` is inferred to be a subclass of `Person`, this tells you there is an error somewhere that you need to track down.

6.5   SUMMARY LEARNING

Syntax, Blank Nodes, and Examples

There is a huge difference between Manchester and Turtle syntax. The former is very easy for humans to read. It achieves this by hiding a lot of what is going on under the covers with triples and blank nodes. If you want to see those things, then Turtle is a good option.

A number of OWL constructs take lists of things as arguments. These include `owl:intersectionOf`, `owl:unionOf`, and `owl:oneOf` (the latter is used for enumerated classes). Lists in Turtle are enclosed in round parentheses.

Filter classes are most often simple class IRIs, but they can be arbitrarily complex class expressions. For example, there must be some collateral for a security agreement, but we don't need a class called `Collateral`, since it is not really a kind of thing, but rather it is something of value that is owned by someone. The filter class in this case is the intersection of two restrictions nested inside another restriction.

Inference

Inference works both ways when you define a class using `:equivalentClass` as in the `:PatientVisit` example. If you know an individual is member of the class, you can infer that it is an event with a care provider and care recipient that are both persons. However, it also goes the other way, if you have an individual and you come to know that it is an event with a care provider and care recipient that are both persons, then you can infer into the class `:PatientVisit`.

OWL is capable of drawing inferences with partial information. Classification inference is useful in a number of ways. It:

- determines new subclass relationships;

- determines which classes an individual belongs to;

- makes it easier to maintain the ontology; and

- helps to point out errors.

This concludes the introduction and detailed exploration of all the major components of OWL. In the next chapter we point out some things you cannot do in OWL and describe some workarounds.

CHAPTER 7

OWL Limitations

In this chapter, we briefly introduce some variants of OWL and describe some limitations of the most widely used variant: OWL 2 DL.

When first released in 2004, there were three OWL variants (sometimes called species): OWL Lite, OWL DL, and OWL Full. When OWL 2 was released in 2012, OWL Lite went away and OWL Full was updated. This chapter will focus on OWL 2 DL, which is the most widely used OWL variant. Unless otherwise specified, when I say OWL or OWL DL, I will be referring to OWL 2 DL.

The Main Limitations

There are a variety of limitations that are important to be aware of in your day to day modeling. As with any designed artifact, the design of the OWL language involved navigating various tradeoffs. The major one follows from a result in theoretical computer science saying that even a small amount of additional expressivity can have a major negative impact on the ability to build efficient inference engines. The OWL designers consciously chose to favor inference performance, giving up some expressive power.

This chapter explains the things that you cannot say in OWL DL that were traded off to get that performance. They relate to the idea of metaclasses, what you can have as the object of a triple, n-ary relationships, rules, dates, and cardinality restrictions. These are listed below.

1. You cannot create true metaclasses.

2. There are some things you cannot do with user-created properties and classes that are possible with classes and properties that are part of the OWL language.

3. There is no way to directly represent relationships between more than two things.

4. Certain kinds of rules are not allowed.

5. You cannot specify just a date, you need so specify the time as well.

6. Not all properties can be used with all kinds of restrictions.

The six items above are about what you cannot say in OWL. There is one more limitation of OWL that we discuss in this chapter.

7. It is difficult to build inference engines that are performant when the ontology is combined with enterprise data. Ironically, OWL let's you say too much.

The remainder of this chapter addresses each of these limitations. Pertinent to the first two items is the distinction between properties and classes that the ontologist creates vs. those that are part of the OWL language. See Table 7.1 for examples.

Table 7.1: User-created concepts vs. OWL constructs

	User-created	OWL Construct
Properties	`:gaveCareTo, :partOf,` `:firstName, :date`	`rdf:type, rdfs:subClassOf,` `owl:unionOf, owl:someValuesFrom`
Classes	`:Party, :PatientVisit,` `:LoanContract, :Device`	`owl:Class, owl:ObjectProperty,` `owl:Restriction, rdfs:Literal`

7.1 METACLASSES

Most of the classes that you will be creating will have member individuals where those individuals are just that, individuals. A member of the class `:Person` is an individual person. A member of the class `:Agreement` is an individual agreement. It does not make any sense to think of an individual person or agreement as a set that also has members.

However, this is not always the case. Suppose you want to create a class that corresponds to a product model like iPhone7 or Lenovo T470P. You could create a class called `ProductModel`. It could have two subclasses, `SmartPhoneModel` and `LaptopModel`, with `_iPhone7` and `_LenovoT470P` as member individuals, respectively. In this case, it *does* make sense to think of the members of the `ProductModel` class as more than just individuals. In the world, a particular smartphone model or a laptop model corresponds to the set of all individual phones or laptops that are instances of those particular models. Because they are sets, one might reasonably create classes for each product model.

The fact that their member individuals are themselves classes makes the product model classes quite different from the other classes we have considered. This introduces the concept of a metaclass: a class whose members are also classes.[25] Unfortunately, OWL DL does not allow something to be *both* an individual *and* a class at the same time. Another way to put this is there cannot be two consecutive `rdf:type` links connecting *user-created* individuals and classes. This is depicted in Figure 7.1.

---

[25] In UML, these are called "power types."

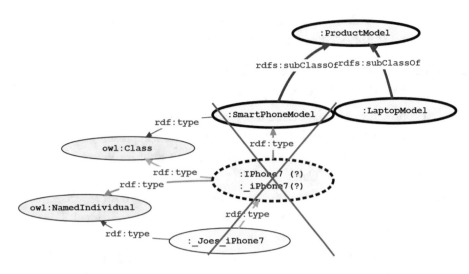

Figure 7.1: Metaclasses are not allowed.

If metaclasses were allowed, our naming convention would not always work. The item would need to both *have* a leading underscore and to *not* have a leading underscore. The brown arrows show the iPhone 7 being represented as an instance of `SmartPhoneModel`. The blue arrows show it as a class whose instances are individual iPhone 7's. The situation is exactly analogous for laptop models.

You have to decide whether you represent the product model as a class or as an instance. Figure 7.2 shows two options for representing the iPhone 7.

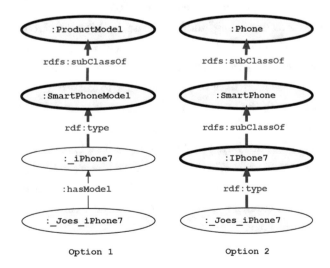

Figure 7.2: Two workarounds to avoid a metaclass.

1. Create the *individual,* `_iPhone7` as an instance of a class, `SmartPhoneModel`. Use the property `hasModel` to indicate the model of a particular phone.

2. Create the *class,* `IPhone7`, in a class hierarchy of kinds of phones. Use the property `rdf:type` to indicate the model of a particular phone.

The first approach would make sense in a product development or marketing context, where there is no interested in tracking individual phones. The second approach would make sense for a retail operation where serial numbers for individual phones matter for inventory and for servicing maintenance plans such as Apple Care.

What if you want to build and integrate systems that handle both aspects? This means you need to say things about the product model, so it will need to be an individual. This is option 1 in Figure 7.2.

What you give up is the ability to use OWL's built in sub-classing and inheritance mechanism for kind of products. For example, because iPhone7 and iPhone would both be instances, you cannot use `rdfs:subClassOf` to relate them. You would have to use a different property, perhaps `skos:broader` and use SPARQL to achieve any inheritance inferencing you may need.

A Non-workaround: OWL Punning

The OWL spec allows you to use a single IRI such as `:IPhone7` to represent both the *class* of iPhones that has individual phones as instances, and also as an *individual* representing a particular model of smartphone. This is called punning, because one IRI is being used to represent two different things.

This contrasts from our approach of using two IRIs, as depicted in Figure 7.2. The OWL spec mentions the words "metamodeling" and "metaclass" in this context, but I find that misleading. Punning only gives the illusion of the true sense of a metaclass. Why? Because the OWL 2 DL semantics dictates that the reasoners treat "different uses of the same name as completely separate."[26] Under the covers, OWL reasoners have to do extra work to keep track of the fact that there are really two different things with the same name. The net result is no different than if you had used two different IRIs. That is why I call it a non-workaround.

You might wonder what this buys you. I wonder too. It's a bit of syntactic trickery that brings to mind the phrase "worse than useless." First, it does not address the fundamental limitation about metaclasses. Second, by giving the illusion of representing metaclasses, it is going to cause confusion. If you manage to avoid confusing yourself, you will almost certainly confuse others trying to understand and use your ontology.

---

[26]   https://www.w3.org/TR/owl2-new-features/#F12:_Punning.

7.2   THE OBJECT OF A TRIPLE

All of the triples we have shown that have user-created properties as the predicate, have objects that are either literals or OWL individuals. However, sometimes it would be convenient to have the object be a class or a property.

Classes as Property Values

Suppose we created the classes `IPhone7` and `LenovoT470P` that we described above. Say we are also interested in classifying books and newspaper and magazine articles by topic, and one of the more narrow topics is the iPhone 7. It would be convenient to just have a triple that connects an article to the class.

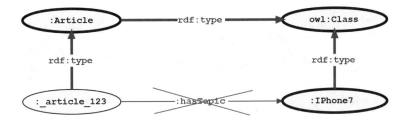

Figure 7.3: The object of a triple cannot be a class.

Unfortunately, this is not possible in OWL DL. We have seen many cases where triples have classes as the object. However, this is only possible when the property already exists in the OWL language, e.g., `rdfs:range`, `rdfs:subClassOf`, and `rdf:type`. You cannot *create* a property and use it in a triple where the object of the triple is a class. This is depicted in Figure 7.3.

This problem and various workarounds are described in detail in the document: "Representing Classes As Property Values on the Semantic Web."[27] A fascinating paper on what it means to be a subject that touches on this issue is "A Formal Ontology of Subject," by Chris Welty.[28]

Properties as Property Values

The object of a triple cannot be a property either; it must be an individual or a literal. The desire for the latter arises when modeling product specifications. But first let's consider a particular phone. Say it weighs 138 g and has 2GB of RAM. An easy thing to do would be to create two data properties called, say `:weightInGrams` and `:gigOfRAM` each with a range of `xsd:decimal`. Then the values for those data properties for that particular phone would be 138 and 2, respectively.

[27]   https://www.w3.org/TR/swbp-classes-as-values/.
[28]   https://pdfs.semanticscholar.org/b93f/5ed4d27056f6a7517589b2f0c63530e1b4e0.pdf.

But, those values come from the technical specifications and should be the same for *every* iPhone 7. The values for specific phones is important for manufacturing tolerances, but not for the purpose of marketing and selling iPhones. For that, you care more about the technical specifications, which often consist of property-value pairs such as the following that apply to the iPhone 7:

1. weight is 138 g; and

2. GB of RAM is 2.

The technical specification for a product consists of several such entries, each, in turn consisting of a property and a value. You could have instances corresponding to each of these entries. Each instance would link to a property and a value, as depicted in Figure 7.4. The problem with this is that you cannot have a property as the object of a triple that has a user-created property as the predicate.

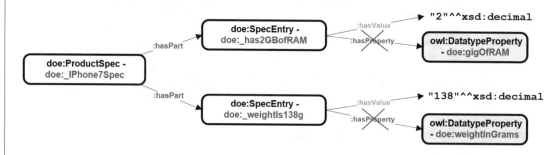

Figure 7.4: The object of a triple cannot be a property.

The workaround is similar to what we did in option 1 of Figure 7.2. There, we represented what would have been the class of iPhone 7s as an individual that corresponded to the iPhone 7 product mode. Similarly, we will not model the properties indicating RAM or weight as properties in OWL, but rather they will be individuals. It would be confusing to create a class called "Property," so, instead, we create the class `Characteristic` which means the same thing. We create a class called `SpecEntry` to represent characteristic-value pairs and we create a class called `ProductSpec` to represent detailed specifications of products, mainly consisting of instances of `SpecEntry`. We'll use `hasPart` to link the product specifications to their spec entries and `hasCharacteristic` to link the spec entries to their characteristics. Figure 7.5 shows the ontology and the updated triples. Figure 7.6 shows the Turtle syntax. Although this example has just three classes and two properties, the core idea illustrated has been used at more than one company, and it has gone into production.

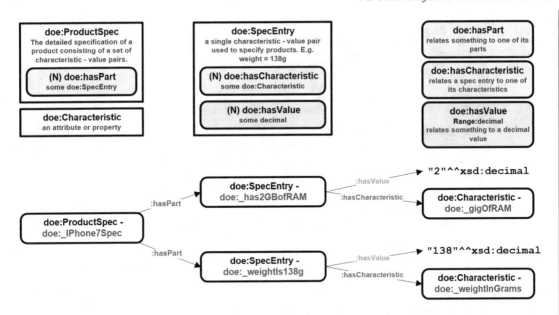

Figure 7.5: Representing a property as an individual.

There are many additional things to take into account in a production system. For a start, if you needed a wide variety of units, having properties like "weightInGrams" would result in a proliferation of similar properties where the unit is buried in the name of the property. In this case, you would want a richer ontology for quantities, units, and measures that could represent conversions and perhaps support dimensional analysis. Also, the specification for a given characteristic is not always a simple number. More generally it is a constraint on what the values must, may or may not be. For example:

1. it might be a range (e.g., 2–5 volts);

2. it might be a minimum (battery life) or a maximum (amperage); or

3. the values might be qualitative rather than quantitative (red, blue, green, or a scale like low, medium, and high).

Having It Both Ways

We have shown workarounds that enable you to make progress, but it's always a choice. In the iPhone example, depending on whether you were more concerned with the phone models or the individual phones, you would represent the concept of an iPhone 7 with an instance or as a class. What if you need both?

```
@prefix : <http://ontologies.demystifyingowl.com/examples/meta#> .
@prefix doe: <http://ontologies.demystifyingowl.com/examples#> .
@prefix owl: <http://www.w3.org/2002/07/owl#> .
@prefix rdf: <http://www.w3.org/1999/02/22-rdf-syntax-ns#> .
@prefix xml: <http://www.w3.org/XML/1998/namespace> .
@prefix xsd: <http://www.w3.org/2001/XMLSchema#> .
@prefix gist: <http://ontologies.semanticarts.com/gist#> .
@prefix rdfs: <http://www.w3.org/2000/01/rdf-schema#> .
@base <http://ontologies.demystifyingowl.com/examples/meta> .

<http://ontologies.demystifyingowl.com/examples/meta>
                              rdf:type owl:Ontology  .

#    Object Properties
doe:hasCharacteristic rdf:type owl:ObjectProperty ;
                         rdfs:comment "relates a spec entry to one of its
                                 characteristics " .
doe:hasPart rdf:type owl:ObjectProperty ;
            rdfs:comment "relates something to one of its parts" .

#    Data properties
doe:hasValue rdf:type owl:DatatypeProperty ;
             rdfs:comment "relates something to a decimal value" ;
             rdfs:range xsd:decimal .

#    Classes
doe:Characteristic rdf:type owl:Class ;
                   rdfs:comment "an attribute or property" .
doe:ProductSpec rdf:type owl:Class ;
                rdfs:subClassOf [ rdf:type owl:Restriction ;
                                  owl:onProperty doe:hasPart ;
                                  owl:someValuesFrom doe:SpecEntry
                                ] ;
                rdfs:comment "The detailed specification of a product consisting
                   of a set of characteristic - value pairs." .

doe:SpecEntry rdf:type owl:Class ;
              rdfs:subClassOf [ rdf:type owl:Restriction ;
                                owl:onProperty doe:hasCharacteristic ;
                                owl:someValuesFrom doe:Characteristic
                              ] ,
                              [ rdf:type owl:Restriction ;
                                owl:onProperty doe:hasValue ;
                                owl:someValuesFrom xsd:decimal
                              ] ;
              rdfs:comment "a single characteristic - value pair used to specify
                 products. E.g. weight = 138g" .

#    Individuals
doe:_gigOfRAM rdf:type doe:Characteristic .
doe:_weightInGrams rdf:type doe:Characteristic .

doe:_has2gigOfRAM rdf:type doe:SpecEntry ;
                  doe:hasValue 2 ;
                  doe:hasCharacteristic doe:_gigOfRAM .
doe:_weightIs138g rdf:type doe:SpecEntry ;
                  doe:hasValue 138 ;
                  doe:hasCharacteristic doe:_weightInGrams .

doe:IPhone7Spec rdf:type doe:ProductSpec ;
                doe:hasPart doe:_has2gigOfRAM ,
                            doe:_weightIs138g .
```

Figure 7.6: Technical specifications in simplified Turtle.

In the case of the properties for technical specification. The solution described will work, so long as the main concern is specifications about characteristics such as weight and not the ability to represent the weight of individual things. If we want to represent specifications as we have done, but also assign weights to individual units, then we have to create a property for weight. We now have an instance and a property to model weight with no good way to connect them.

One workaround is to make use of the fact that an annotation property can have any IRI as a value. So when you would otherwise want to use an object property with a class or property as a value, use an annotation property instead. You will lose out on the ability to use OWL DL reasoners, but you can create and link the data you need as triples and access it in an application using SPARQL.

7.3 N-ARY RELATIONS

Consider how we might represent the following statements.

1. Michael Uschold was employed at Boeing from 1997–2008.

2. Stearyl alcohol is used as a surfactant in a shampoo.

3. The capacity for Emirates Stadium for the 2015–16 season is 60,362.

In the first case, we can represent the underlying fact that Boeing employs Michael as a simple triple using an employs property as follows: `:_Boeing :employs :_MichaelUschold`. However, if we need to say that this employment relationship held for a period of time, no single triple will work.

In the second case, there are three things of interest: a material, how the material is being used and a product. Unlike the employment example, there is no underlying relationship between two things that is being qualified. Also, you cannot just say stearyl alcohol is used as a surfactant, because it is only used that way in some products, it plays other roles in other products. Similarly, you cannot just say that stearyl alcohol is used in shampoo because many shampoos use other surfactants. Again, no single triple will work. This is a three-way relationship.

In the third case, there are also three things, the venue, the season, and the capacity. Capacity is not an attribute of the venue on its own, nor is it an attribute of the season. Conceptually, it is an attribute of the combination of the venue and the season. This is necessary because seats may be added or removed from season to season. Again, there is no single triple that can capture this.

Recall that an OWL property represents a way that two things can be related. This is also called a binary relation. But none of the above relationships can be captured by an OWL property because more than two things need to be related. The more general case of a binary relation is called an n-ary relation where n can be more than just 2.

If OWL provided direct support for n-ary relations, we might represent the above examples as follows. Each item in the parentheses is commonly referred to as an argument.

1. employment (MichaelUschold, Boeing, 1997, 2008);

2. materialUsageForProduct (Stearyl alcohol, surfactant, shampoo); and

3. venueCapacityForSeason (Emirates, 2015–16, 60,362).

But OWL does not provide such support. Fortunately there is a standard workaround for these examples. The idea is to turn the n-ary relations into classes whose instances correspond to each of the above facts. This means you are turning a particular relationship into a thing. This is sometimes called "reification."[29] This is a more general case of what is called "RDF reification"[30] which turns an RDF triple into a thing. We are not talking about that here.

How does it work? You create a class that has restrictions pointing to each of the n arguments. With n arguments, there will be n triples and n+1 instances (the main instance and one for each argument) (see Figure 7.7). Each rounded shape is an individual, whose class IRI is in a bold black font. Beneath that is the individual's IRI in blue. The green rounded shapes in the **Employment** instance are data property assertions whose values are literals, not other individuals. They each represent a triple that is not explicitly shown as such.

The employment example is a special case in which there is an underlying binary relation which can be defined as a property chain and be inferred (hence the dotted line in figure).

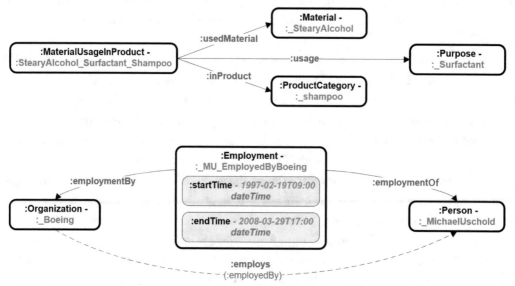

Figure 7.7: Reifying n-ary relations.

[29] https://en.wikipedia.org/wiki/Reification_(knowledge_representation).

[30] https://www.w3.org/TR/rdf-primer/#reification.

Note that it is often difficult to find good natural-sounding names for the properties for such instances. That is because the instances are often not natural objects, they are modeling artifacts created to work around the binary relation limitation. There is no everyday word or phrase to refer to a "`MaterialUsageInProduct`" nor a "`VenueCapacityForSeason`." In the case of employment, it's not so bad, people do work for companies for periods of time and we may refer to them, e.g., "my employment at Boeing."

7.4   RULES

Recall from Section 3.1.1 that an inference rule specifies what new information can be logically concluded from a set of premises. For example for any two classes, `C1` and `C2`, declaring that `C1` is a subclass of `C2` tells the inference engine to use the following inference rule when performing reasoning:

For any individual, x

IF: `x rdf:type C1.`
THEN: `x rdf:type C2.`

So, declaring that `:Lawyer rdfs:subClassOf :Person` tells the inference engine to use the following rule:

IF: `x rdf:type :Lawyer.`
THEN: `x rdf:type :Person.`

We have introduced many such rules to specify the semantics of various OWL constructs, in this case, `rdfs:subClassOf`.

In any complex subject, whether it is healthcare, finance, or manufacturing, there is a wide variety of rules that one can think of where the conclusion logically follows from the premises. However, given the tradeoff between expressivity and inference performance, the OWL designers limited the kinds of rules that one can state. There is one very important broad category of rules that cannot be expressed in OWL DL. As an example, we will try to generalize the ideas about internal organizations and transactions considered in the previous chapter.

Requiring Two Individuals To Be Different

In Section 6.3, we defined the class `FinancialTransaction` to be an event with a buyer and a seller. If we want to ensure that the buyer and seller are different, we can make the properties `hasSeller` and `hasBuyer` disjoint. This is one case where we can require two individuals to be different, but as we will soon see, it is a special case.

Suppose we wanted to create a class called `InternationalFinancialTransaction` that required the countries that the buyer and seller were based in to be different. We start by asserting that `InternationalFinancialTransaction` is a subclass of the following three classes:

1. a financial transaction;

2. a buying party that is based in some country; and

3. a selling party that is based in some country.

Figure 7.8 shows what this looks like in Protégé, which is using Manchester syntax.

Figure 7.8: International financial transaction.

This is another example of a chained restriction. We saw this when we modeled security agreements in Section 6.2. This is fine as far as it goes, but it says nothing about the countries being different from each other. It turns out that you cannot say this in OWL DL.

There are different ways to approach this. However you proceed, it is a good idea to add a comment into the definition of international financial transaction that says something like: "NOTE: OWL DL cannot express the requirement that the countries are different." This successfully communicates to any human using the ontology how that class is expected to be used. Unfortunately, it does not allow us to infer a given financial transaction to be an international one. Nor will it be possible to catch an error if someone erroneously creates an explicit instance of an international financial transaction when both the buyer and seller are based in the same country.

The only way to accomplish these things is to extend or augment OWL with features that support a wider range of inference rules. Several such languages have been proposed, including SWRL,[31] SPIN,[32] and more recently SHACL,[33] which is based on SPIN. It is also possible to use a SPARQL CONSTRUCT query to assert that a given individual is a member of the `:InternationalFinancialTransaction` class.

---

[31] https://www.w3.org/Submission/SWRL/.
[32] https://www.w3.org/Submission/spin-overview/.
[33] https://www.w3.org/TR/shacl/.

The details of rule languages are beyond the scope of this book. We will show how this might work using a SWRL-like syntax. SWRL has been around for a long time, and Protégé has a plugin for it, so you can experiment to get some inferences to work. We use the following notation:

- `C(?x)` is short for the triple: `?x rdf:type C.`

- `p(?x,?y)` is short for the triple: `?x p ?y.`

- `<=>` means logically equivalent to

```
InternationalFinancialTransaction(?t) <=>

    FinancialTransaction(?t) and
    hasBuyer(?t,?buyer) and Party(?buyer) and
    hasSeller(?t,?seller) and Party(?seller) and
    basedIn(?buyer,?country1) and
    basedIn(?seller,?country2) and
    differentFrom(?country1,?country2)
```

Thus, if you know that `?t` is an international financial transaction, then you can conclude that there are buyer and seller parties based in two different countries. Conversely, if you know that there are buyer and seller parties based in two different countries, you can conclude that `?t` is an international financial transaction. This works just like the `owl:equivalentClass` construct.

Even if this was integrated into a logic reasoner there is nothing to stop someone from putting in the same bad data. But if you asserted an individual to be an instance of `:InternationalFinancialTransaction` and the two countries were the same, the inference engine would find a contradiction and notify you.

Requiring Two Individuals To Be the Same

The flip side of an international financial transaction is a domestic one. That means that the buyer and seller are based in the same country. The same underlying limitation of OWL DL that prevents one from fully expressing the meaning of an international financial transaction makes it impossible to fully express the meaning of a domestic financial transaction. The SWRL-like rule above can be tweaked to cover this case by renaming the class to be `DomesticFinancialTransaction` and changing the `differentFrom` to `sameAs`. This would work, but it would be easier to just use a single variable for the country. This obviates the need to use `sameAs`. See below:

```
DomesticFinancialTransaction(?t) <=>

    FinancialTransaction(?t) and
```

```
hasBuyer(?t,?buyer) and

hasSeller(?t,?seller) and

basedIn(?buyer,?country) and

basedIn(?seller,?country)
```

A similar situation arises if we want to capture the meaning of an internal transaction, where the buyer and seller are both from the same organization. In Section 6.3, we defined the class :InternalSemviaTransaction, and successfully inferred an individual to be one of its instances.

This worked fine, but it was from the perspective of a single organization, Semvia. Suppose we wanted to capture the general idea of an internal financial transaction. We want to say that both the buyer and seller are parts of the *same* organization without specifying which organization that might be. This also cannot be said in OWL DL, but as in the prior examples, we can say it in a more general rule language. For example:

```
InternalFinancialTransaction(?t) <=>

    FinancialTransaction(?t)

    hasBuyer(?t,?buyer) and   Party(?buyer) and

    hasSeller(?t,?seller) and Party(?seller) and

    partOf(?buyer,?org) and

    partOf(?seller,?org) and

    Organization(?org).
```

How To Tell When You Need a Rule that OWL Does Not Support

If you get into modeling in a serious way, you will inevitably encounter situations like this. You might think you can do it in OWL, and spend hours trying, only to fail. You may be none the wiser about whether the limitation was in your own creativity or the expressivity of the OWL language. A good rule of thumb, as it were, is to watch for situations where coming up with an accurate English definition requires you to use words that mean "different from" or "same as," as in the above examples. If so, chances are you cannot say it in OWL. Another good resource to use is Stack Overflow. [34]

A Point on Terminology

There is some ambiguity in how the term "rule" is used in the context of OWL. People will often say OWL cannot do "rules." But as we said above there are many dozens of inference rules that OWL

---

[34] https://stackoverflow.com/questions/tagged/owl.

does support. This apparent contradiction is resolved by realizing that the word is being used in two different senses, a broad sense and a narrow sense. In the narrow sense of the term, it is true that OWL does not support "rules." What this means is that you cannot do the kinds of things we just talked about in this section—where you need to ensure sameness or differentness in certain ways in an expression to convey the meaning you want. In technical terms, you cannot use a variable to ensure co-reference as we did with `?country` when defining a domestic financial transaction and with `?org` when we defined an internal financial transaction.

7.5 DATES AND TIMES

In the spirit of reuse, the developers of OWL borrowed from a large set of existing XML Schema datatypes rather than inventing something new. Not everything is available in OWL, but the smaller set is more manageable.[35] One case where a genuine inconvenience arises is for dates and times. Where XML gives a moderate variety of options for specifying dates and times, OWL basically gives only one: `xsd:dateTime`. It represents a specific date and time which is a concatenation of the date and the time plus an optional time zone offset from Universal Coordinated Time (UTC). See Table 7.2 for a few examples.[36] If you are not going to use a time zone offset, then make sure it is clear from context.

Table 7.2: Examples of `xsd:dateTime` literals	
`"1977-09-29T09:30:00"^^xsd:dateTime`	9:30am, September 29, 1977
`"1977-09-29T13:30:00Z"^^xsd:dateTime`	1:30pm on March 29, 1977 UTC
`"1977-09-29T09:30:00-4:00"^^xsd:dateTime`	9:30am on Sept. 29, 1977 US Eastern Daylight Time

Specifying a Date Only

This means that you cannot just specify a date without specifying a time. This can be very inconvenient. We describe three ways to deal with this.

1. Stay in OWL DL, and use a programmatic workaround.

2. Stay in OWL DL and use or create a custom date and time ontology.

3. Go outside of OWL DL and use `xsd:date`.

The first option entails living with the inconvenience, and using some convention to indicate that you only care about the date and not the time. For example, you can treat any `dateTime` literal

---

[35] See: https://www.w3.org/TR/owl2-syntax/#Datatype_Maps for details.
[36] At this date and time I got on my bicycle and pedaled from Buffalo, NY to San Diego, CA.

whose time part is `00:00:00` to be just the date. If you needed to specify the very beginning of the day, you could use `00:00:01` to include the intended meaning of `00:00:00`, which is now reserved by convention for when no time is wanted. You would have to enforce this programmatically, when you write application code or SPARQL. This is workable, and allows you to take full advantage of OWL 2 inference.

The second option is to create your own classes and properties to provide the support you need. This approach was taken by the developers of the financial industry business ontology (FIBO[37]). Alternatively, you can find an existing ontology that does this for you. Be careful about importing an existing ontology; it might be overly complicated. Be equally careful about getting too involved in creating lots of features that you might not need.

The third approach is to step out of OWL DL, staying in RDF and just use `xsd:date` and `xsd:time` from XML Schema. When you use this approach, some of the OWL DL reasoners will not run. This limits your ability to check the consistency of the ontology during development and evolution.

I almost always go with the first option. The complexity of option two is rarely needed, and I like to check my ontology with an OWL DL reasoner.

Date Math

Another thing that causes problems is the inability to add a duration to a date and get another date. First, there is nothing in OWL for the idea of a duration (e.g., 1 second, 12 hours, 3 weeks). So you will need to create or borrow an existing ontology that defines a duration. This will likely take you into the general idea of units and measures.[38] SPARQL 1.1 supports `xsd:date + xsd:yearMonthDuration` and `xsd:dateTime + xsd:dayTimeDuration`, and other things. It is also possible to convert `dateTime` literals to numeric form (e.g., as in Unix time[39] which uses the number of seconds since January 1, 1970) and do the math after that.

7.6 CARDINALITY RESTRICTIONS WITH TRANSITIVE PROPERTIES OR PROPERTY CHAINS

Suppose you want to represent a subsidiary as an organization that is part of another organization. You might create the classes and properties in Figure 7.9. We create the class, `Subsidiary` stating that it is a subclass of `Organization` as well as the restriction "`partOf min 1 Organization`." Thus, if we know something is a subsidiary, we know it is an organization that is part of

---

[37] https://www.edmcouncil.org/financialbusiness.

[38] For a simple, yet expressive, ontology for units and measures including durations, see: https://semanticarts.com/gist.

[39] https://en.wikipedia.org/wiki/Unix_time (accessed January 23, 2018).

another organization. But not all organizations that are part of other organizations are subsidiaries. For one, it would have to also be a legal entity.

We have seen a number of examples of cardinality restrictions. We have seen that properties can have one or more of various characteristics, such as functional, symmetric, and transitive. We reuse the transitive `partOf` relationship that can connect many different kinds of things. This looks just like many other examples we have seen, and you would have no reason to expect a problem. However, if you try to run a description logic inference engine on an ontology with this small ontology, you may bump up against another limitation of OWL DL.

You get different behaviors from different inference engines. The error messages are all different, and not always very helpful. At the time of this writing, the inference engine Hermit 1.3.8.413 does not complain about this particular issue—it is not clear what it is doing. Pellet is fussier. It gives an error message, which is an important clue. It says "An error occurred during reasoning… `UnsupportedFeatureException`" and names the property: "`TransitiveObjectProperty(doe:partOf)`." This tells you *where* the problem is but not *what* the problem is.

For obscure technical reasons that are well outside the scope of this book, it turns out that a transitive property may not be used with a cardinality restriction, it takes you out of OWL DL. In this case, there is a simple workaround, replace the restriction "`partOf min 1 Organization`" with "`partOf some Organization`." Logically they mean the same thing, but the inference engines treat cardinality differently.

The `min 1` case is the special case where there is an easy workaround—not so if you want a min cardinality with a number other than 1, or if you want any max or exact cardinality.

If we use the restriction with `exactly 1` instead of `min 1`, both Pellet and Hermit complain. Pellet gives more or less the same error message as before, naming the offending property and saying there is an unsupported feature exception.

A more readable version of what Hermit literally says is: "The non-simple property `partOf` or its inverse appears in the cardinality restriction "`partOf max 1 Organization`." Notice that there is a `max`. That is because `exactly 1` is just a convenient shorthand for two separate restrictions, a `min 1` and a `max 1`.

```
doe:partOf rdf:type owl:ObjectProperty ,
                    owl:TransitiveProperty ;
           rdfs:comment "relates something to that which it is a part of" .

doe:Organization rdf:type owl:Class ;
                 rdfs:comment "a social entity created to pursue one or
                     more goals, typically having persons as members" .

doe:Subsidiary rdf:type owl:Class ;
               rdfs:subClassOf doe:Organization ,
                               [ rdf:type owl:Restriction ;
                                 owl:onProperty doe:partOf ;
                                 owl:onClass doe:Organization ;
                                 owl:minQualifiedCardinality "1"^^
                                     xsd:nonNegativeInteger
                               ] .
```

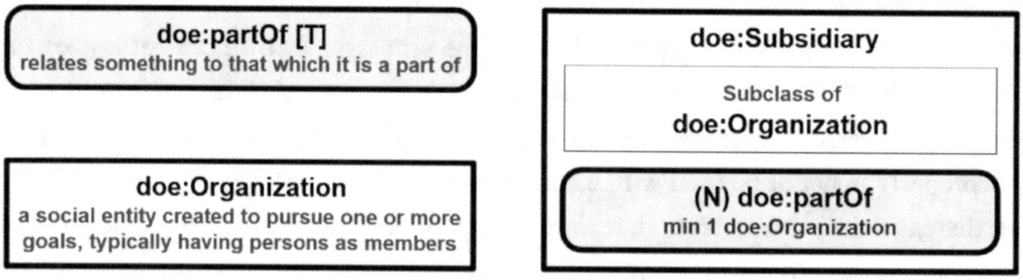

doe:partOf [T]
relates something to that which it is a part of

doe:Organization
a social entity created to pursue one or more goals, typically having persons as members

doe:Subsidiary
Subclass of **doe:Organization**
(N) doe:partOf min 1 doe:Organization

Figure 7.9: Cardinality restriction with transitive property

For the purpose of building OWL ontologies, it is sufficient to know that non-simple properties typically involve transitive properties and/or property chains. If you are interested in the details, have a look at the OWL spec.[40] Be warned, it is not for the faint of heart.

In summary, when creating a cardinality restriction (i.e., one that uses min, max or exactly) you should avoid using a transitive property or a property defined using a property chain.

Workarounds

The simple case, as noted above is to never use min 1, always use some. They are logically equivalent, and the reasoners work better with the latter. If you want to use a cardinality restriction with n greater than 1 with a transitive property or a property chain, then you need to give something up. The situation is analogous to that for rules. We want to express true facts about the subject, but OWL will not let us. Again, the task is to decide what to give up and what to keep.

For min cardinality, the easiest thing is to just use an owl:someValuesFrom restriction that means min 1, but to have a text annotation saying what the minimum really is. For example, an agreement between two parties has a minimum of two obligations. But if you use the transitive

[40] https://www.w3.org/TR/owl2-syntax/#Global_Restrictions_on_Axioms_in_OWL_2_DL.

property, `:partOf`, you cannot say that. So you can use "`partOf some Obligation`" instead, and the text comment should indicate that there should be two obligations.

An alternative is to keep the cardinality part intact, but to make sure the property in the restriction is neither a property chain nor transitive. One way to do this is to keep the transitive property but to not use it in that restriction. Create a non-transitive version that means the same thing in the world. This solution is workable, but care must be taken to avoid confusion and mistakes due to their being two properties that mean essentially the same thing.

You might be tempted to just ignore the inference engine. That will solve one problem, but is likely to leave errors in the ontology that the inference engine would have caught. Another strategy is to use a less fussy inference engine that will catch many logical errors, but will not complain about some of the more subtle details. An alternative is to separate out the parts of the ontology that are not OWL DL compliant, and run the inference engine on the rest. That is fiddly, but can also be made to work.

In any event, the things that are true in the world that you are not expressing formally due to OWL limitations should be documented in annotations. This enables users of the ontology to know what is needed by way of added constraints or specialized application code is needed to honor the intended semantics.

Property Chains

As noted above, you cannot use a property chain with a cardinality restriction. I no longer use OWL property chains. If I need a property chain, I just use SPARQL.

7.7 INFERENCE AT SCALE

Recall from Section 3.1.1 that OWL is a subset of the more powerful first-order logic. The particular subset was chosen to hit a sweet spot between computational efficiency for inference, and representational expressiveness. However, in practice for large triple stores with data, this sweet spot is not sweet enough. Performance and scalability is limited.

From the beginning, there was a concern for inference performance. The first version of OWL was released in 2004. There were three versions: OWL Lite, OWL DL, and OWL Full with increasing expressive power and decreasing inference performance. A key difference that OWL Full brought was the ability to represent true metaclasses. The most widely used original version was OWL DL.

Based on user feedback, a new version called OWL 2 was released in 2012. OWL Lite went away, OWL Full was updated, but, to my knowledge, neither version of OWL Full ever got much traction. Efficient reasoners were never developed.

Today, the term OWL DL usually refers to OWL 2 DL. OWL 2 added a variety of new features, but what they are is now just a matter of historical interest.[41] What is important is the in-

---

[41] https://www.w3.org/TR/2012/REC-owl2-overview-20121211/#Relationship_to_OWL_1.

troduction of three new sub-languages of OWL 2 called profiles, each designed to meet a different need. They are: OWL 2 QL, OWL 2 EL, and OWL 2 RL.[42] Each limits the expressive features available in different ways in order to improve specific kinds of reasoning.

For Large TBoxes: OWL 2 EL

Many large biomedical ontologies have large numbers of classes and properties defined using a variety of complex expressions. The need is to infer additional relationships, and there is minimal focus on instance data. The EL profile is the preferred choice for this situation. Think "E" for expressive, EL is the most expressive profile; it also rhymes with "T" for TBox.

For Large ABoxes: OWL 2 QL

If you need to run inference over large amounts of instance data, and you want to provide database applications with an ontological data access layer, OWL 2 QL is a good choice. Its expressivity has been restricted so that any query over an OWL 2 QL data set can be expressed in SQL. Think 'Q' for SQL.

For Rules: OWL 2 RL

If you live in a business rules context and want traditional business rule engines to run efficiently, then OWL 2 RL is a good choice. The RL profile has also become the preferred approach for reasoning with Web data. Think "R" for rules.

All of the axioms can be expressed as logical implications, which is a synonym for a broad interpretation of the term "rule." As we discussed in Section 7.4, OWL cannot express rules at all, in the narrower interpretation of the term "rule."

7.8   SUMMARY LEARNING

Due to the design constraints of OWL, there are certain things you cannot do. These relate to the idea of metaclasses, what you can have as an object of a triple, rules, property chains, dates, and cardinality restrictions.

Metaclasses and Objects of Triples

You cannot use a class or property as the object of a triple with a user-created property. OWL DL does not allow you to create something that is both an individual and a class. You have to decide which. OWL 2 DL provides punning which only gives the illusion of having something be a class

---

[42]   http://korrekt.org/papers/Kroetzsch_OWL_2_Profiles_Reasoning_Web_2012.pdf.

and an individual. This limitation is related to not allowing metaclasses. To be precise, you cannot create an individual, `X` and two classes `C1` and `C2` where: `X rdf:type C1` and `C1 rdf:type C2`. This would make `C2` a metaclass, which is a class whose members are themselves classes.

Although OWL users cannot create metaclasses, OWL itself does have metaclasses. For example, `_Michael rdf:type Person` and `Person rdf:type owl:Class`. The metaclass is `owl:Class`.

N-ary Relations and Reification

All variants of OWL are limited to binary relations. Every triple has a subject and an object. The need commonly arises to represent relationships that hold between more than two individuals. Reification allows you to get the effect of an n-ary relationship using binary predicates only.

Rules

Many inference rules are not expressible in OWL. Specifically, if you require two variables to be the same or different, you probably cannot say it in OWL. Although OWL has many inference rules built in, there is a narrower sense of the term "rule" that is limited to more expressive rules that OWL cannot represent.

Dates and Times

OWL does not have a datatype that allows you to specify a date without also specifying a time. Neither OWL nor SPARQL directly support adding a duration to a date and getting a new date out of the box.

Cardinality Restrictions with Transitive Properties or Property Chains

Cardinality restrictions are not allowed with transitive properties or property chains. Always use an `owl:someValuesFrom` restriction instead of a `min 1`. When you know something is true but cannot represent it in OWL, it is good practice to indicate that in a text annotation.

This limitation will not affect your ability to load triples into a triple store, but a DL inference engine may well complain. One strategy is to use a less fussy inference engine.

Inference at Scale

OWL 2 provides three profiles to improve inference at scale, at the price of decreased expressivity. OWL 2 EL is for large TBox reasoning; OWL 2 QL is for large ABoxes and OWL 2 RL is for rules.

CHAPTER 8

Go Forth and Ontologize

This final chapter offers a variety of practical tips and guidelines to put the wrapping touches on what you need to know to get out there and build some real-world ontologies. We'll have a look at some modeling tools, principles, patterns, and pitfalls to speed you along your way.

Although the sections in this chapter have been thoughtfully ordered, there are relatively few dependencies among them. Feel free to skip around.

8.1   MODELING PRINCIPLES AND TOOLS

8.1.1   CONCEPTUAL AND OPERATIONAL

In traditional relational data modeling there are three distinct modeling levels: conceptual, logical, and physical. The first captures the semantics of the subject matter of the data but is not executable and therefore is not operational in any implementation. The physical model is executable and operational. Unfortunately, the process of going from conceptual to physical involves various optimizations and some of the semantics is lost. It was always a bit of a fairy tale that the logical and conceptual models would be kept in synch as the database evolved. It just didn't happen.

This problem does not arise when you create triple store databases using ontologies as the schema. The ontology is loaded directly into a triple store and used in an operational system. Therefore it is both conceptual and operational.

Keep this in mind when designing your ontology. It's a balancing act. You don't want to just put more and more into your ontology because it is cool and seems important. Be guided by what you expect triples to look like in triple stores of various applications that will use the ontology. But don't go so far as to include a lot of highly application-specific things that are not really about the subject being modeled.[43]

8.1.2   CONCEPTS, TERMS, AND NAMING CONVENTIONS

The saying "a rose by any other name is still a rose" is just as true in logic as it is in English or any other natural language. Although it makes no difference to the computer or the inference engine, choosing good names makes it easier for the human to understand and use the ontology. So choose names carefully.

---

[43]   For more information on conceptual vs. operational ontologies, see: https://www.slideshare.net/UscholdM/conceptual-vs-operational-a-false-distinction

The word "term" is commonly used to refer to a name of a concept. The informal term will typically be the label on the concept. The formal term will be the IRI, usually abbreviated by just using the local name. It can also have alternate labels to represent synonyms.

In principle, because OWL does not make the unique name assumption, a single concept can have more than one IRI. In practice that will not arise unless you are interacting with ontologies built by others.

Terms are important, but not as important as the concepts. In healthcare, a key concept is that of being a person who receives care during a patient visit, the term that names that concept is `doe:Patient`. By analogy, the word "rose" is the term that is used to name the concept we all know about in our world, described in detail in Wikipedia,[44] and other places.

Ideally, when a person sees a term, the meaning you intend is the first one that comes to mind. Avoid ambiguous terms. For example, one definition of "loan" is: "an amount of money loaned at interest by a bank to a borrower, for a certain period of time."[45] Here, the loan is the amount of money. But the money is really just a part of the broader agreement between the borrowing and lending parties with a repayment schedule and other contract terms. Use of the ambiguous term, "loan" could lead to wrong or inconsistent use of the ontology and should be avoided. Focus first on the key concept, and then give it a name. Here, the loan contract is key, and a good class name would be: `LoanContract`.

Avoiding ambiguous names often leads to longer names (e.g., `LoanContract` vs. `Loan`). You can use the OWL annotation property, `rdfs:label` to provide a prettier name, e.g., "loan contract" for use in the user interface of software driven by the ontology.

Don't Let Terms Get in the Way

When modeling, it often happens that terms can get in the way. If you are focused on coming up with an OWL representation of the term "loan" and there are multiple parties that must agree, you may quickly run into trouble because people use the term differently. The same is true for the term "risk." For example, in the sentence "there are several risks if we go off-piste skiing in the mountains today", the word "risk" refers to something bad happening. If you are in the mountains and stop to consider the avalanche risk if you ski down a steep gully, then the word "risk" is a synonym for probability.

Different meanings will come to mind for different people when they see terms like "loan" and "risk." Don't risk using such overloaded terms (so to speak).[46]

---

[44]   See https://en.wikipedia.org/wiki/Rose (accessed on January 16, 2018).
[45]   See http://www.dictionary.com/browse/bank-loan (accessed on January 16, 2018).
[46]   See "The Importance of Distinguishing between Terms and Concepts," https://semanticarts.com/blog/terms-concepts-whats-important-pt-1/,

Naming Conventions

Other than blank nodes, resources are named with IRIs. The main things you will be naming are classes, properties, and individuals. However, each ontology also has its own IRI. You will also be naming the ontology files. It is good practice to name the ontology file to exactly echo the local name of the ontology IRI. For example, if the ontology IRI is `http://ontology.myco.com/event`, then the main part of the file name should be "event." Think twice before including the file extension in the IRI because the same ontology can be saved out in various syntaxes or file extensions.

The following conventions for naming classes and properties are widely used.

1. Classes: singular noun phrases in upper camel case,

   e.g., `LoanContract`

2. Properties: tense neutral verbs in lower camel case,

   e.g., `hasBorrower`

There are no standard conventions for naming individuals. You can use the one we use in this book (prefixing with an underscore). Or you may wish to look around and choose among other conventions, or invent your own.

It is critically important that every IRI is globally unique. An important place to start is to use only those namespaces whose domains you control. For example, at Semantic Arts, `http://ontologies.semanticarts.com/gist#Person` is the IRI for the person class.

Generally, you should follow any conventions that are widely used independently from your organization. This makes it easier for others to understand your ontology. Only change things up if you need to, and document the decisions and rationale.

Meaningless IRIs

Although the practice of IRIs being suggestive of the meaning of the resource is nearly universal, there are compelling reasons to have the IRIs themselves be meaningless, and to use labels and other annotations to convey meaning to the humans. For example, instead of `doe:DataSystem`, you would have an IRI such as `doe:UffcsEE3452X` and the label would say: "data system."

Why would anyone do this? Suppose someone decides that the term "data system" is misleading and what's really meant is "data store." If the IRI is meaningless, you can leave it as it is and just change the label. There is minimal impact.

If the IRI was `doe:DataSystem`, then you have an ontology update problem. All the triples using the old IRI need to change. All the SPARQL queries referring to the globally unique IRI in applications using that ontology need to change.

It's a tradeoff. The problem with meaningless IRIs is that you never have any idea what they mean by looking at them. There always has to be some kind of tool support to show you the meaningful labels.

Summary: Naming Conventions

1. Ensure that all IRIs are globally unique and resolvable as URLs.

2. Follow widely used conventions when possible. Otherwise document your decisions and apply them consistently across your organization.

3. Use IRIs and labels that are suggestive of the meaning of the concept. Err on the side of long and specific names to avoid ambiguity.

4. Think about the tradeoffs of using meaningless IRIs and see what makes sense in your situation.

8.1.3   MODELING CHOICE: DATA OR OBJECT PROPERTY?

Always use an object property when you are relating two individuals with IRIs. Consider a data property when it is natural to think of the property as having a value, such as a string, date, or number. Most of the time it is pretty clear what to do. If you are not sure, ask the following question:

Will you need to say something about the value of the property?

If not, you can safely choose to use a data property. If so, then you will need an object property. See Section 4.8.2 for a detailed analysis of when to use datatype vs. object properties.

8.1.4   MODELING CHOICE: CLASS OR PROPERTY?

Most of the time it is clear whether a concept should be modeled as a class or as a property. Use a class to represent a kind of thing that is conveniently described using a noun phrase. Use a property to represent a relationship that is conveniently described using a verb phrase. For each of the terms in Table 8.1, decide whether you think you should represent the underlying concept as class or as a property.

Table 8.1: Classes or properties?

patient	event	owns	has first name	is CEO of
governs	currency	symptom	loan contract	corporation
employs	occupies	bank	occurred at	legal entity

Many of these are fairly straightforward. However, it is not always so obvious, especially when roles are involved. You could create an `employs` object property connecting an organization to the persons that it employs. If you need to track additional information about employment such as start and end dates, then you can use a class to model the concept behind the word "employs" that appears in the table. See Figures 7.7 and 8.8 for a standard way to model employment.

As another example, consider the borrowing and lending parties on a loan contract. Below are three ways to model this (see Figure 8.2).

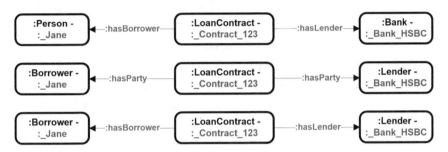

Figure 8.2: Three ways to model borrowing and lending.

Using properties: You could have `hasBorrower` and `hasLender` as properties that link the contract to the borrowing and lending parties, (either persons or organizations). In this case, the concepts of borrowing and lending are represented as properties.

Using classes: You could have a more general property called `hasParty` that links to instances of classes `Borrower` and `Lender`. Here the concepts of borrowing and lending are represented as classes.

Using properties and classes: You can combine the two and represent the concepts of borrowing and lending both as classes and properties. Then the contract would be linked to the instances of `Borrower` and `Lender` using the properties `hasBorrower` and `hasLender`, respectively.

The questions below provide criteria for choosing a good approach.

1. Which approach introduces the fewest new things to cover the same requirements? Avoid redundancy; less is more.

2. Is the class a "true" kind of thing, or do individuals just happen to be members of the class as a result of being in a relationship with other things?

3. Are you representing more meaning than you need to?

Ignoring `Bank`, `Person`, and `hasParty`, which are not specific to borrowing, the first approach introduces two properties, the second approach introduces two classes, and the third introduces all four. It needlessly represents the semantics of borrowing both as classes and properties and thus scores low by the first criteria.

Representing borrowing as a class scores low on the second criteria because being a borrower is just a consequence of being the borrowing party on a loan. Lending is slightly different, because some lenders lend as a business. So far so good, by the second criteria. However, what about all the persons and organizations that are lenders only because they are the lending party on a loan? So, having a class called Lender would be confusing. Does it refer to only organizations that are inherently lenders, or to any organization or person that happens to lend? Therefore, the second approach does not get high marks for the second criteria.

The third criteria is about meaning. Is there any meaning that needs to be expressed on a special borrower class that cannot be carried by a property relating the contract to the person? The answer depends on the circumstances. But chances are you don't need it. When in doubt, leave it out.

8.1.5   MODELING CHOICE: CLASS OR INDIVIDUAL?

The question of whether to model something as a class or an individual usually does not arise. Model it as a class if it is a kind of a thing which has instances. Model it as an individual if it has no instances. Again, it is not always so obvious. For example, Starbucks is an individual corporation. Is each Starbucks store an instance of Starbucks? Think for a moment about this. Should we consider representing Starbucks the corporation as a class with individuals?

The answer boils down to the nature of the relationship between Starbucks the corporation and an individual Starbucks store. Is the store an instance of Starbucks? If not, what is it? One thing we can say is that Starbucks the corporation owns each store. So what is each store an instance of? We could create a class called `StarbucksStore` which is a subclass of `Store` which is necessarily owned by Starbucks the corporation. It is now more clear that a specific person like you or me and the specific corporation Starbucks should be represented as individuals, not classes. Individual people or corporations are not kinds of things having instances.

In Section 7.1, we saw that a product model such as the iPhone can be viewed *both* a member of some class, and also as a kind of thing that has individual members. The right choice depends on your specific needs (see Figure 7.2).

8.1.6   MODULARITY FOR REUSABILITY

Any ontology that will cover a useful part of your enterprise will have a mix of topics, some more specific than others. It is useful to create separate ontologies for different topics, and to have the more specific ontologies import the more generic ones. For example, most of the product specification ontology we built for electrical products had nothing to with electrical products. We took great care to split out anything that was specific to electrical products into a separate ontology. That way the product specification ontology could be used for other products across a range of industries.

In another example, suppose your business includes tracking various kinds of entertainment including sports, cinema, and theater. The central concept is an event that people can attend, whether it is called a performance, a game, or a showing.

Let's say you start with theater and you will be tracking performances, theater groups, venues, and attendance. You might start to put classes in the theater ontology for things like performance, venue, capacity, ticket sales, etc. However, most of this has nothing to do with theater per se. Venues are important for a wide range of events, outside of the entertainment arena. Venues involve geography and the people and/or organizations that operate them. Entertainment always involves people and organizations that are putting on and/or attending the events. One way to organize this subject matter into ontology modules is depicted in Figure 8.1.

As discussed in Section 3.1.11, the construct for importing is `owl:imports`. The result of ontology A importing ontology B is the union of the sets of triples for both ontologies.

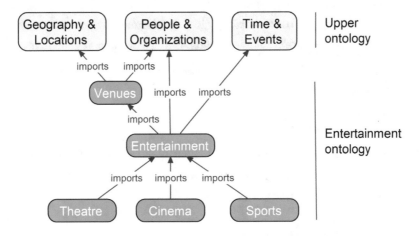

Figure 8.1: Ontology modules for entertainment.

8.1.7   ONTOLOGY EDITORS AND INFERENCE ENGINES

There are a variety of tools for building ontologies. The most commonly used are Protégé and Top-Braid Composer. The former is open source. The latter is a commercial tool that has a free version. It's a good idea to try each to see what you prefer. Both can be used to build OWL ontologies, covering most or all of what you are likely to need.

TopBraid Composer offers more robust support for RDF and SPARQL than Protégé, but the latter offers better support for OWL inference. Protégé has a variety of OWL DL reasoners built in and others are available as plugins. TopBraid Composer provides inferencing for some of the OWL 2 profiles, but none that will support all of OWL 2 DL.

Because OWL is a standard, either tool can load the output of the other. Therefore, you can choose to work with both tools, depending on your needs. There is a minor caveat: TopBraid Composer does not support files with a .owl extension. Use .rdf or .ttl instead.

Available inference engines include Hermit, Fact++, Pellet, and TrOWL. Each of Pellet, Hermit, and Fact++ behave differently—some go faster on some ontologies but slower on others. They support OWL 2 DL, but run slowly on large complex ontologies. TrOWL goes dramatically faster (from minutes to seconds). This is because it leaves out inferences regarding datatypes.

If inference still takes too long, it is time to look at the ontology to see what you can change to speed things up. It is a bit of a black art. If you have lots of instances, try removing them. Sometimes too many inverse properties or disjoints used in certain ways can be a problem. In the worst case, it is a matter of trial and error.

Be careful about vendors claiming that they have ontology development tools that support OWL. The support can vary widely. Find out if it was developed natively to support OWL or whether OWL as an add-on. Find out what if any loss of information occurs when importing OWL 2 ontologies, and whether it supports round-tripping. Otherwise, once it gets in, it can't get out!

8.2   MODELING PATTERNS

8.2.1   GENUS-DIFFERENTIA

Many of the classes we have defined use what is known as the genus-differentia style of definition. The genus is the existing class on which the new class is based, the differentia says what is different about the new class. This is good practice (see Table 8.2).

Table 8.2: Genus-differentia style definition

Class Being Defined	Genus	Differentia
Patient	Person	Received care on a patient visit
PatientVisit	Event	There is a care provider and a care recipient; both are persons
SecurityAgreement	WrittenContract	Has an owned thing with estimated value as collateral
SemviaOrganization	Organization	Is part of the overall Semvia organization
InternalSemviaTransaction	FinancialTransaction	Both the buyer and seller are Semvia organizations

8.2.2   ORPHAN CLASSES AND HIGH-LEVEL DISJOINTS

At the end of Section 2.3.5, we showed how stating that two classes are disjoint can help spot errors. This illustrated an important and useful pattern. Namely, to have a relatively small number of high level classes, and to carefully specify the disjointness relationships that exist between them.

Look at the inconsistent ontology example in Figure 8.3. The problem is that the individual, `AdmitPatient` is a member of two disjoint classes.

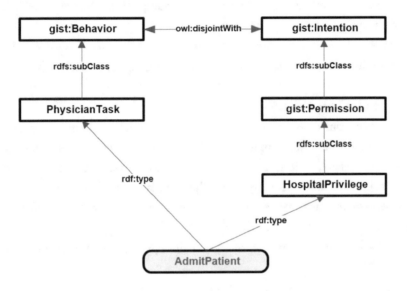

Figure 8.3: High-level disjoints.

The highest-level classes are orphans, i.e., they have no parent class. In our simple example, there are five classes, and just two orphans. You want a small number because it is far easier to use the high-level disjoints pattern. If, instead of the five-classes depicted in Figure 8.3, the hierarchy was completely flat, you would need to make ten decisions, one for each pair of classes. With just two high-level classes, we only need to make one. Furthermore, this single decision results in 5 inferred disjointness axioms. Saying that `Behavior` is disjoint from `Intention` allows us to infer that:

```
:PhysicianTask        owl:disjointWith    :HospitalPrivilege,
                                          :Permission,
                                          :Intention.
:HospitalPrivilege    owl:disjointWith    :PhysicianTask,
                                          :Behavior.
```

A realistic ontology will likely have dozens or possibly hundreds of classes. Take gist,[47] for example, as depicted in Figure 6.14. This version has 131 classes and just 26 orphan classes in the asserted hierarchy, which reduces the number of decisions from 8,515[48] in the case of a totally flat hierarchy to 325. After inference, there are just 18 orphan classes and only 153 pairs of classes to consider. That is manageable.

The next question is how best to ensure that there are a small number of orphan classes? First, you should make liberal use of the genus-differentia pattern. The new class is inferred to be the subclass of the genus class.

Often it is easy to do this, but sometimes it is trickier. Ask yourself: what exactly is this thing? One way to make this task easier is to connect to an existing upper ontology. That is the next topic.

8.2.3   UPPER ONTOLOGIES

It is useful to connect your ontology to an upper ontology, which has generic concepts that are not specific to any subject, industry, or application. There are a number of advantages. First, you don't have to reinvent the wheel, so it saves a lot of time. Second, you can build a better ontology by starting with a solid foundation and leveraging the axioms to find errors. Find one that has been developed by people you trust over a long period of time. Many of the wrinkles will have been worked out and it will be more stable and reliable.

What to look for in an upper ontology? In the enterprise context, look for the following characteristics:

1. The upper ontology is easy to learn, understand, and use. This means manageable scope, being well documented and structured with plenty of axioms to specify the semantics.

2. It matches your way of thinking and modeling your enterprise.

3. It is mature and relatively stable, yet is still evolving and supported.

4. It is suitable for business use, designed by business people for business people. An academically first rate ontology may not be a good fit for your enterprise.

5. There is a community of users.

A comprehensive review of existing upper ontologies is beyond the scope of this book.[49] A reasonable fit to the above ideal is gist. It is the only upper ontology specifically designed for the enterprise. It is free and open source, so long as you give appropriate attribution.

---

[47]   https://www.semanticarts.com/gist/.

[48]   If there are n classes, then there are (n*(n-1))/2 different pairs of classes.

[49]   There was a recent meeting devoted to the topic: http://ontologforum.org/index.php/SummerInstitute2017.

Non-business users will have different requirements. For example, there is an upper ontology that is widely used by scientists called BFO—Basic Formal Ontology.[50]

A note on terminology. The notion of "upper" is somewhat relative. The most useful way to think about it is whether an ontology has many of the higher level concepts that you can reuse. If you want an ontology for enterprise applications, you can think of gist as an "upper enterprise ontology."

How an Upper Ontology Helps

The following deceptively simple example illustrates some of the benefits of importing an upper ontology. Suppose you are tracking data centers, each having unique identifiers. One straightforward way to model this is to create the two classes `DataCenter` and `DataCenterID` that are linked by the property `hasIdentifier` (with inverse `identifierFor`). A restriction requires every `DataCenter` to have a `DataCenterID`. Figure 8.4 shows what this looks like in Protégé.

Careful inspection of this model will reveal two problems. One illustrates a common mistake that arises when using OWL restrictions. The other is a mistake of omission: an important characteristic about identifiers is left unsaid. Neither of these two mistakes will be found by running inference, and they could impact downstream applications.

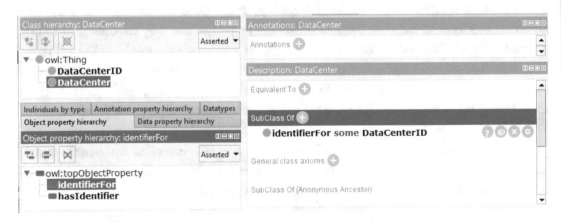

Figure 8.4: Simple ontology for data centers.

We will now show how we can catch both of these errors by importing and mapping to a simple upper ontology like gist, which already models identifiers. The `DataCenterID` class is a subclass of `gist:ID` and `DataCenter` is a subclass of `gist:Building`. The property `gist:identifies` (with inverse `gist:identifiedBy`) could be used directly, but the datacenter ontology already has different IRIs for the same properties as we saw above. The property `identifierFor` is equivalent to `gist:identifies`, and their respective inverses are also equivalent.

[50] http://ifomis.uni-saarland.de/bfo/.

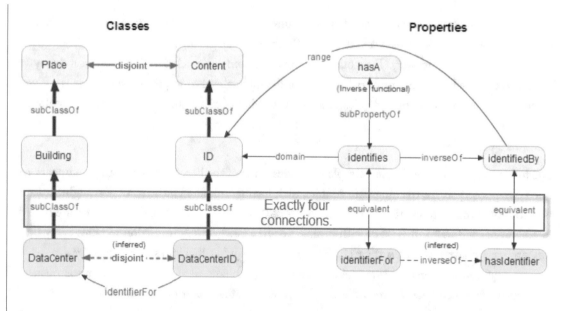

Figure 8.5: Connecting to an upper ontology.

We say as much using owl:equivalentProperty, as described in Section 4.9. The data center properties now have all the characteristics of the gist properties that they are equivalent to (e.g., inverse functional). This is depicted in Figure 8.5. Figure 8.6 shows what this looks like in Protégé.

If we run inference, DataCenter becomes equivalent to owl:Nothing (which represents the empty set). That means DataCenter cannot possibly have any members. This is clearly an error. To find the mistake, you click on the question mark to get an explanation justifying the inference (see Figure 8.7).

The explanation is not very human-friendly, but it does offer clues. The root cause is that the restriction used the property the wrong way around (identifierFor instead of hasIdentifier). A more human-friendly explanation for how the error was caught is as follows.

1. A DataCenter is a gist:Building which is, in turn, a gist:Place.

2. The range of gist:identifiedBy (which is equivalent to identifierFor) is gist:ID.

3. The restriction says that the DataCenter is the identifierFor at least one DataCenterID.

4. Because identifierFor is equivalent to gist:identifies, which is the inverse of identifiedBy, it has a domain of gist:ID.

5. Therefore, DataCenter is inferred to be a gist:ID.

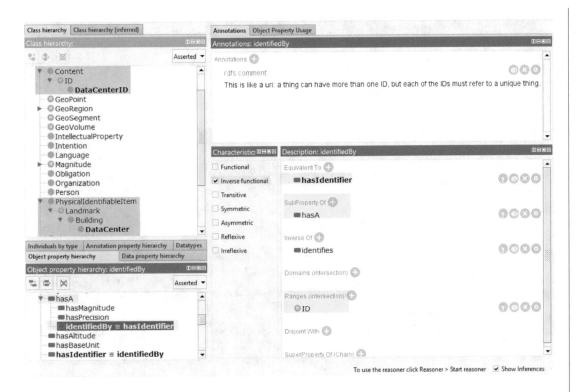

Figure 8.6: Data centers in Protégé.

6. But gist:ID is disjoint with gist:Place.

7. Therefore, DataCenter is both a gist:Place and a gist:ID, but those two classes are disjoint, so DataCenter cannot have any members.

If you explicitly added a DataCenter instance, the ontology would become inconsistent, which is a worse problem than one class not being able to have any members. The mistake of omission is corrected automatically by connecting up the properties. An identifier must identify only one thing. This characteristic is inherited from the gist:hasA property which is inverse functional. In summary, by making a few simple connections to an existing upper ontology, the following benefits are possible.

1. The existing model for identifier includes things that the modeler of the new ontology might or might not have thought of (e.g., disjoint classes, property characteristics).

2. There is no need to model the concept of an identifier, it is already done for you.

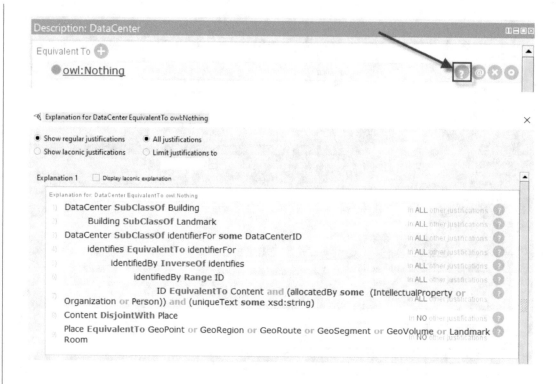

Figure 8.7: Detecting and resolving bugs.

3. The inference catches logical inconsistencies, such as the common mistake of getting the property direction wrong on an OWL restriction.

4. The accuracy and completeness of the ontology is improved, resulting in fewer mistakes in downstream applications.

5. To the extent that other ontologies will be built in a similar manner, the problem of silos is much less likely to arise. The ontologies will have a shared core.

6. It will be possible to query across multiple data stores without doing any additional mapping. For example, if someone is interested in getting a list of buildings in a given area, they may be surprised to see data centers turning up. This happened because `DataCenter` is a subclass of `Building.` The effort to connect to the upper ontology is modest compared to the work to map across silos after the fact.

In short, using an upper ontology helps you build better ontologies faster. It takes less time because you can reuse already modeled concepts. Your ontology is better because by hitching a ride on the semantics of the imported ontology, inference can catch errors that you can easily miss.

8.2.4 N-ARY RELATIONS

As described in Section 7.3, OWL properties directly represent binary (2-ary) relations. If you want to represent relationships between 3 or more things (n-ary), you need a workaround. Figure 7.7 shows how OWL is used to represent these n-ary relations. N is 3 for the material usage example (the material, the purpose, and the product) and n is 4 for the employment example (the employer, the employed, the start date, and the end date).

The general pattern for reifying an n-ary relationship involves creating a class with n restrictions. An instance of that class representing the relationship holding between n particular individuals requires n+1 triples—one to create the instance, and one for each restriction (see Figure 8.8).

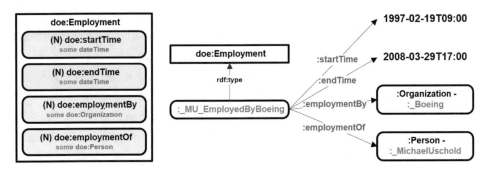

Figure 8.8: Representing a 4-ary relationship with 5 triples.

8.2.5 BUCKETS, BUCKETS EVERYWHERE

As categorizing machines, we humans like to create metaphorical buckets, and put things in them. Different kinds of buckets are modeled differently in OWL. The most common bucket represents a kind of thing, such as person, or building. Bill Gates goes into the person bucket, and the Edinburgh Castle (Figure 3.2) goes into the building bucket. Such buckets are represented as OWL classes and we use `rdf:type` to put things into the bucket. For example:

`:Person rdf:type  owl:Class.`

`:_BillGates rdf:type :Person.`

Another kind of bucket is for when you have a collection of things, like people on a jury, that are functionally connected in some way. Those related things go into the bucket. For this case, we create a class called `Collection` and a subclass called `Jury` whose instances represent the buckets containing individual jurors, e.g., the jury for the OJ Simpson trial. Use `isMemberOf` to put the jurors into the bucket. For example:

```
:Collection          rdf:type          owl:Class.
:Jury                rdfs:subClassOf   :Collection.
:_Jury_OJ            rdf:type          :Jury.
:_CarrieBess         :isMemberOf       :_Jury_OJ.
:_SheliaWoods        :isMemberOf       :_Jury_OJ.
:_LionelCryer        :isMemberOf       :_Jury_OJ.
```

Convince yourself that a collection does not represent a kind of thing. A jury is a kind of thing, a particular jury is not.

A third kind of bucket corresponds to a tag which represents a topic and is used to categorize individual items for the purpose of indexing a body of content. For example, the tag "_Winter" might be used to index photographs, books and/or YouTube videos. Any content item that depicts or relates to winter in some way should be categorized using this tag.

The representation for this echoes how we represent collections. The differences are (1) the bucket is an instance of a subclass of `Category`, rather than of `Collection` and (2) we put things into the bucket using `categorizedBy` rather than `isMemberOf`.

The tag for winter corresponds to a bucket containing all the things that have been indexed/categorized using that tag. Some of the classes and properties used here are borrowed from gist.

Table 8.3: Different kinds of buckets			
Kind of Bucket	**Example**	**Representing the Bucket**	**Putting something in the Bucket**
Individual of a Kind	Bill Gates is a Person	Instance of `owl:Class`	`rdf:type`
A bucket containing functionally connected things	Sheila Woods is a member of OJ's Jury	Instance of a subclass of `Collection`	`isMemberOf`
A bucket containing all the items with a particular tag	The book "Winter of our Discontent" has the tag, "Winter"	Instance of a `Tag`, a subclass of `Category`	`categorizedBy`

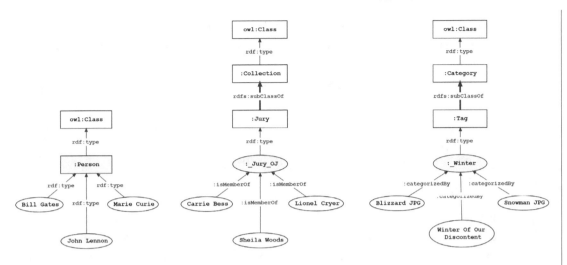

Figure 8.9: Representing different kinds of buckets.

8.2.6 ROLES

Table 8.4: Using properties to represent roles

Class	Role property	Role property	Role property
:PatientVisit	:careProvider	:careRecipient	
:LoanContract	:hasBorrower	:hasLender	:hasOriginator
:PropertyAppraisal	:hasAppraiser	:hasSeniorAppraiser	:hasProperty

Roles are fairly pervasive in modeling. They arise most frequently for events, agreements, and when reifying n-ary relations. For example, the event of a patient visit always has one person as the patient and at least one as the care provider. There may be others participating in some capacity, say a physician's assistant. The event of playing any game will have at least two individuals or teams playing a competitor role. In some cases, those roles are distinguished, e.g., home team vs. away team.

Roles also show up in agreements. There are always at least two parties to any agreement. For example, an employment contract has an employer and an employee. A software licensing agreement has a licensor and the licensee. A loan contract has a borrower and a lender. There is an important event attached to a mortgage contract, which is a property appraisal, which in turn has other roles, e.g., the appraiser (a person) and the property being appraised.

Earlier, in Section 8.1.6, we discussed different ways to model such situations. There is a simple approach that works most of the time illustrated at the top of Figure 8.2. It is to represent the semantics of the role in a property, rather than create classes for each role. The main advantage of this approach is that it is simple and it works (see Table 8.4).

It's easy to make this overly complicated, and even easier to get confused—I've been there. Stick with this simple approach unless you have specific examples that won't work. I know exactly what might justify a more complex approach, but in my many years of building enterprise ontologies in many industries, it almost never arises.

This completes our discussion of modeling patterns. Next, we consider some common pitfalls.

8.3   COMMON PITFALLS

8.3.1   READING TOO MUCH INTO IRIS AND LABELS

A term that is suggestive of its meaning helps humans understand the intended meaning. However, it is easy to fall into the trap of thinking that the computer will somehow be able to divine that information also. The computer "knows" only as much as you tell it. For further discussion on this, see Section 3.1.8.

8.3.2   UNIQUE NAME ASSUMPTION

If you are used to systems that make the unique name assumption, it may catch you off guard that in OWL two different IRIs can both refer to the same thing.

If an individual is linked by a functional property such as `hasBiologicalMother` to two different IRIs, then a triple connecting the two IRIs using `owl:sameAs` can be inferred (see Figure 4.13).

8.3.3   NAMESPACE PROLIFERATION

It is a common practice to have a different namespace for every ontology. We regard this as a bad practice. Why? It makes refactoring the ontology more difficult and much more disruptive. If you need to move some classes or properties from one ontology to another, you have to change the IRI. This means you have to change every use of that IRI. That could impact hundreds of triples in dozens of ontology modules for large ontologies with many separate ontology modules.

That is time-consuming and error-prone if done by hand. Automation is possible but may be disruptive if there are multiple ontology authors working on different modules.

There is also a risk of disruption to the user community. They will have to change all their data and application code that makes use of the old IRIs. Although this can be mitigated by using the OWL deprecation feature (discussed in Section 8.4.5), having lots of deprecated things around creates clutter. In our experience, it is better to limit the introduction of new namespaces to situations where there is a different governance body that is minting the IRIs.[51]

---

[51]   See: "Finding and Avoiding Bugs in Enterprise Ontologies" for an example: http://ceur-ws.org/Vol-1586/codes1.pdf.

8.3.4   DOMAIN AND RANGE

Given two things connected by a given property, domain, and range are used to indicate what classes those two things *necessarily belong to*.

The most common pitfall is to think of domain and range as integrity constraints, they are not. Instead, domain and range sanction certain inferences. If you get surprising inferences that don't make any sense, double check that you are using domain and range correctly.

Another common pitfall is not realizing that if you specify two or more domain classes, then the actual domain of the property is the intersection of those two or more classes. The same goes for multiple range classes.

Finally, be careful not to define the domain or range too narrowly. This limits the possibility of reuse. Additional details on pitfalls of using domain and range are found in Section 4.4.1.

8.3.5   LESS IS MORE

It is easy to be lured into thinking that more is better. It's not. Smaller and simpler ontologies are easier to understand and use, and more computationally efficient as well. A good principle to use is "when in doubt, leave it out."

One thing to be wary of is building out a nice big class hierarchy for something when most of the distinctions will never be used. Below we consider two specific aspects of the "less is more" principle. One is to keep the number of primitives small. The other is to avoid proliferation of properties.

Create and Use a Small Number of Primitives

The ontology will be easier to understand and use if the number of primitives is relatively small and other concepts are built up from these primitives. For example, consider the definitions of `Patient`, `PerformedProcedure`, and `Party`. Each is defined entirely from existing lower level concepts. The genus-differentia pattern is used for `Patient`. These examples are illustrated in Figure 8.10. A "primitive" is where you stop. You don't define it in terms of other things, it is foundational.

Figure 8.10: Class equivalence reusing existing concepts.

Avoid Proliferation of Properties

One way we keep the number of primitives small is to avoid introducing new properties that don't add new semantics. In the genealogy domain, say we have the property, hasSibling and we want to model brothers and sisters. One way is to create two sub-properties, hasBrother and has-Sister, whose ranges are Male and Female, respectively. See the top of Figure 8.11. Note that the dotted green arrows show what can be inferred.

Alternatively, we can capture the semantics of brother and sister using a restriction with the inverse of the existing property, hasSibling. This way we can define the brother and sister concepts entirely in terms of existing primitives with the same number of classes and without creating any new properties.

The first approach captures the essential difference in meaning between Brother and Sister by creating two new properties and using Male and Female as their domains, and then using those properties in restrictions. The second approach expresses the exact same meaning by reusing the hasSibling property as part of the genus-differentia pattern where the genus classes are Male and Female and the hasSibling property is used in the restrictions expressing the differentia. (see Figure 8.11).

Keeping the number of primitives low has practical value. The fewer things you have, the easier it is to find what you need. It helps during ontology development and evolution.

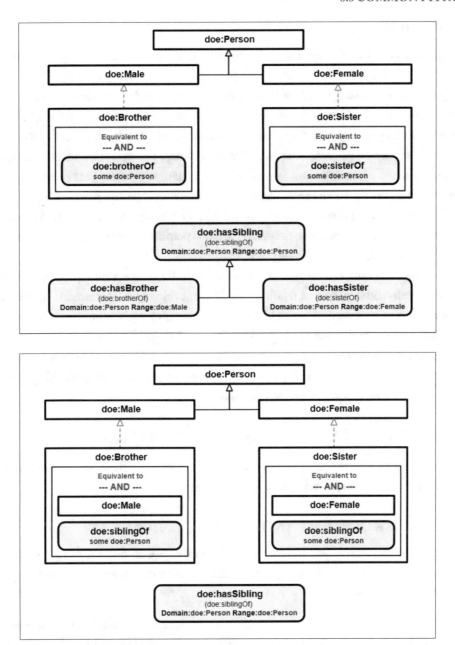

Figure 8.11: Avoiding property proliferation.

Exercise 1: What justifies the inference of `Brother` into `Male` and `Sister` into `Female` in the examples shown in Figure 8.11.

When To Create New Properties?

We have seen a number of examples analogous to the sibling, brother, sister example. You can use `hasPart` for many circumstances, rather than numerous variations such as `hasWheel` and `hasChapter`. There are many kinds of identifiers including vehicle identification numbers and serial numbers. You don't need separate properties to connect to each kind, e.g., `hasVIN`, `hasSerialNumber`. Rather, you can represent the different kinds of identifiers as classes (`VIN`, `SerialNumber`) and use a single property such as `isIdentifiedBy` to connect things to their identifiers.

Sometimes the opposite occurs. Instead of having a single property that points to several different classes, there are times when you want several different properties pointing to the same kind of thing. For example, the properties `hasAnnualSalary`, `hasCreditLimit`, and `hasPrice` all point to an amount of money. These properties are not pointing to different kinds of money, they denote different relationships that things have with money.

The common theme here is that it makes sense to introduce a new property when it represents semantically different way of relating two different things, and you need to know about that difference to support your requirements. If the new subproperty you wish to create is only different because it has a narrower domain or range, there is a good chance you don't need it.

At times, if a group has been using particular terms describing relationships for a long time, it can be better to just have some extra properties, even if they are not adding anything semantically. It can make the ontology easier to understand and use.

8.4 LESS FREQUENTLY USED OWL CONSTRUCTS

So far, we have focused on the most widely used OWL constructs. There are some additional ones that we briefly mention here, in case you may need them.

8.4.1 PAIRWISE DISJOINT AND DISJOINT UNION

Pairwise disjoint: If you have a set of classes that are all disjoint from one another, you can use the `owl:AllDisjointClasses` construct to avoid the need to specify each pair, one by one,

Disjoint union: If you want to partition a class into subclasses that are pairwise disjoint and their union is equal to the starting class, then you can use the `owl:disjointUnionOf` construct.

8.4.2   DATATYPES

At the end of Section 3.1.5, we went over the most commonly used datatypes, however there are many others. The complete list may be found in the OWL2 Reference Card[52]; they are fully documented in the OWL 2 Specification.[53] There are various kinds of numbers (`real`, `float`, `positiveInteger`) as well as some miscellaneous datatypes such as `Boolean`, `hexBinary`, and `base64Binary`.

It is also possible to create custom datatypes. For example, you could define a datatype and call it SSN for social security number and require it to be 9 digits where the first digit has to be a 0 or 1. Due to poor inference performance, this facility is rarely used.

8.4.3   DIFFERENT INDIVIDUALS

Because the unique name assumption does not hold for OWL, the inference engine does not assume that two different IRIs refer to two different individuals. If you want inferences that will only fire if two individuals are different, then you have to explicitly declare them to be different. For example:

`:_JanSmith  owl:differentFrom :_JanAnnSmith.`

Pairwise different from: You can use the `owl:AllDifferent` construct to say a group of specified individuals are all different from each other. This works just like `owl:AllDisjoint-Classes`.

8.4.4   SAME INDIVIDUALS

Similarly, two IRIs can refer to the same individual. For example, if you are integrating different databases, you might realize that the two individuals `_Barak_H_Obama` and `_BarakObama` are the same individual. You can say this with the following triple:

`:_Barak_H_Obama   owl:sameAs  :_BarakObama.`

When you are building your own ontology, you will generally not need to use `owl:sameAs`. It comes is most handy when you are using other ontologies or data sets. In Section 4.6, we saw how the inference engine can infer `owl:sameAs` when using functional properties (see Figure 4.13).

8.4.5   DEPRECATION

If an ontology evolves in such a way that may cause disruption to a user community, OWL provides a mechanism to handle deprecation. For example, suppose you want to change the name of the property `memberOf` to `isMemberOf`. You would deprecate the former and make it equivalent to the latter. This works by using the Boolean annotation, `owl:deprecated`. For example:

---

[52] https://www.w3.org/2007/OWL/refcardA4.
[53] https://www.w3.org/TR/owl2-syntax/#Datatype_Maps.

```
:memberOf    owl:deprecated           "true".

:memberOf    owl:equivalentProperty   :isMemberOf.
```

This gives the downstream users the option to leave the old property in place and update to the new property if and when desired.

8.5   THE OPEN WORLD REVISITED

As explained in Section 3.1.3, OWL uses open-world inference. This means that the inference engine is open to the possibility that there is *more* to know than what has been directly asserted. In short, there is a difference between "no" and "don't know." Fortunately, you don't have to worry too much about it in the early stages of learning OWL.

Practical Import

What does the open world mean in practice? Most of the time when you use a description logic (DL) reasoner on the TBox to check the consistency of your ontology, the fact of open world reasoning won't be so noticeable. We highlight two situations when you want to be aware of the open world.

1. You cannot infer an individual to be a member of a universal (all values) restriction, nor a max or exact cardinality restriction.

2. You can put lots of restrictions on your classes that you know to be true in the world, even if there is no expectation that any one application will use them all.

Preventing Inference into Restriction Classes

If you are trying to use inference to classify individuals, you need to be aware that you can never infer an individual into a restriction class using max or exact cardinality. Why? Consider the bicycle that must have exactly two wheels. Because of the open world, the inference engine will always allow that there might be other wheels that it does not know about, so it can never be certain that there are only two. It can be certain if there are more than two, in which case it will conclude that the individual is not a member of the class.

You also cannot infer into an all values restriction for a similar reason. There might be another assertion that comes along that breaks the pattern.

Even though you cannot make certain inferences, it can still be a good idea to include these restrictions. Why? Because it helps achieve the goal of faithfully representing subject matter. It communicates to the humans using the ontology what is intended. This reduces the likelihood of errors in applications using the ontology. It can also serve as a formal specification for later implementation of integrity constraints using SPARQL or SHACL.

Restrictions that May Go Unused

If you are creating a loan ontology for a financial institution, the central concept will likely be loan contract. There are dozens of possible things you could say about loan contracts by connecting the contract to other individuals or literals. There will be many applications in the enterprise about loans, and every one of those properties will be used in at least one of the applications, but probably no one application will need to use all of them.

During ontology development, when you attach restrictions to the class for loan contract, you are modeling the real world. So you can say whatever is true about loan contracts that you think you will need in your enterprise. So you might have 20 restrictions. In the open world, this is fine. Every loan contract in the world has that many properties.

If you have a closed-world mentality, you may be hesitant to put so many restrictions on a class, for fear that if you don't have the data, then instances of loan contract might not be allowed. Don't hesitate. Just put those restrictions on, as long as they are in scope for the anticipated uses of the ontology. The idea of "allowing" something to be the triple store is where SHACL comes in. That is where you choose just what subset of properties you wish to use, and their cardinality. [54]

Closing the World

Once you load an ontology into a triple store, populate it with data and use SPARQL to query it, the world is no longer so open. A SPARQL query processor will only see the triples that are in the store. If you are confident in your data quality, you could use SPARQL or SHACL to specify rules to assert new triples that a DL reasoner will not infer due to the open world. For example, if you could infer things to be Bicycles. When you move to creating applications and using SHACL and SPARQL, you are for practical purposes, closing the world.

A way to close the world in a localized way is to use enumerated classes that cannot have any more members than the ones explicitly listed.

8.6   SUMMARY LEARNING

Modeling Principles and Tools

An ontology in OWL can be both conceptual and operational at the same time. There are two main ontology editors in widespread use, Protégé and TopBraid Composer. The former has better support

[54]   See: https://www.slideshare.net/UscholdM/putting-fibo-to-use.

for OWL, the latter has better support for RDF and SPARQL. Pay attention first to the concepts, and then think of good terms. It is wise to follow standard conventions for naming, and to document any departures. It's a good idea to break up large ontologies into smaller reusable modules, using the OWL import mechanism.

Given a property that has some kind of value, represent it as a data property unless you need to say something about that value, in which case, use an object property.

Usually it is clear whether to model something as a class or a property. A common exception is for roles, where either choice can work.

It is also usually obvious whether something should be represented as a class or as an individual. However, due to OWL limitations about metaclasses, you sometimes have to resort using an OWL individual to represent what in the world is a kind of thing.

Modeling Patterns

Use the genus-differentia pattern to keep the number of orphan classes small, making it easier to specify high-level disjoints. When using the genus-differentia pattern, connect to high-level classes in an upper ontology, which has generic concepts that apply across many industries and subjects. An upper ontology helps you build better ontologies faster by reusing already modeled concepts and leveraging existing axioms to detect inconsistencies.

To represent an n-ary relationship, create a class with n restrictions. The standard way to represent a "bucket" in OWL is to use a class whose instances are in that bucket. A collection is another kind of bucket, that is for functionally connected things. A third type of bucket corresponds to the set of items categorized with a particular tag.

There is a simple and effective way to model roles that works most of the time. The semantics of the role is captured in a property such as `hasLender` or `hasCEO`. These properties point to persons or organizations that play the role.

Common Pitfalls

Beware of falling into the trap of thinking that the computer will know what you mean by your terms. It only knows as much as you tell it. There is no unique name assumption in OWL. Domain and range are used to infer what kinds of individuals are the subject or object of a triple. They cannot be used as integrity constraints. It is common bad practice to over-constrain domain and range or properties. This limits their use and drives the creation of unnecessary properties.

Less Frequently Used OWL

There are some convenience constructs for specifying disjointness among multiple classes. There are numerous datatypes that are available for special uses and a mechanism for creating your own datatypes. You can declare two or more individuals to be the same or different from each other. There is a deprecation mechanism for evolving ontologies.

Open World

An OWL inference engine does not assume it knows everything. If it cannot prove something is true or false, it is open to the possibility of there being more information out there that it does not know about yet. Once you move to a triple store and use SPARQL, the world is "closed." A SPARQL query processor cannot distinguish between "no" and "don't know", where as an OWL inference engine can.

8.7   FINAL REMARKS

There are challenges and techniques along the way for learning any skill, whether it is fixing things around the house, kayaking down a raging river, or tracking down cyber criminals. Initially, you will make plenty of mistakes and it will take hours to figure out how to do things for the first time. The process of becoming an expert is about learning from mistakes and broadening the scope of situations that you can comfortably handle. The same is true for building an ontology.

There is a saying: "If there are two ontologist in the room, there will be at least three opinions about how to model a given thing". There are many skilled ontologists whose work I know and respect, who do things differently than I do.

The process of becoming expert also entails having a sense of when and whether to "break the rules". Expert photographers know which scenes work well divided in two, and will successfully break the rule of thirds. In this book I have shared a number of guidelines for what I consider to be good practice. As your expertise grows, you will begin to question some of the recommendations in this book. It is good to evaluate and re-evaluate why you do things in a certain way. If you decide to break a rule, do so with care and understand your rationale. Sometimes it matters less how you do something, as to do things in a consistent way.

GO FORTH AND ONTOLOGIZE!

Appendices

A.1 ACRONYMS & ABBREVIATIONS

ABox: Assertional Box
BFO: Basic Formal Ontology
CIO: Chief Information Officer
DL: Description logic
ER: Entity Relationship
FIBO: Financial Industry Business Ontology
FOL: First-Order Logic
IRI: International Resource Identifier
OO: Object-Oriented
OWL: Web Ontology Language
RDF: Resource Description Language
SPARQL: SPARQL Protocol and RDF Query Language
SHACL: Shapes Constraint Language
SPIN: SPARQL Inferencing Notation
SQL: Structured Query Language
SWRL: Semantic Web Rule Language
TBox: Terminological Box
URI: Uniform Resource Identifier
UTC: Coordinated Universal Time.
W3C: Worldwide Web Consortium
XML: Extensible Markup Language

A.2 E6TOOLS VISUAL OWL SYNTAX

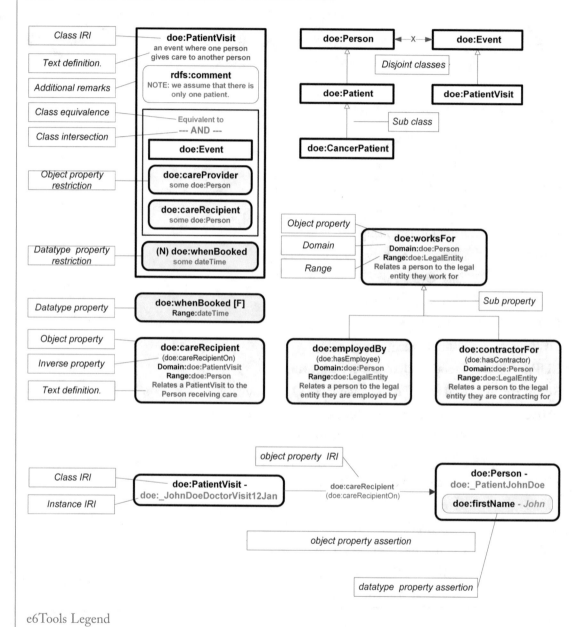

e6Tools Legend

A.3 RECOMMENDED RESOURCES FOR FURTHER LEARNING

To round out your knowledge of RDF, RDF Schema, and OWL, I recommend two books. *The Semantic Web for the Working Ontologist* is closest to this book, having similar goals and target audience. It covers much of the same ground in a very different way. A *Semantic Web Primer* is an academic text in its third edition. It has been translated into dozens of different languages.

Below are selected reference materials that you may find useful, arranged by topic.

Semantic Web Technology Stack

Semantic Web for the Working Ontologist: Effective Modeling in RDFS and OWL, (2nd edition), by D. Allemang and J. Hendler. Morgan Kaufmann, 2011

A Semantic Web Primer, by G. Antoniou, P. Groth, F. van Harmelen, and R. Hoekstra. MIT Press, 2012

Learning SPARQL, by B. DuCharme, Sebastopol, CA: O'Reilly, 2011

Validating RDF Data, by J. E. L. Gayo, E. Prud'hommeaux, I. Boneva, and D. Kontokostas, Synthesis Lectures on the Semantic Web: Theory and Technology, Morgan & Claypool, 2018. DOI: 10.2200/S00786ED1V01Y201707WBE016.

"A Description Logic Primer," https://arxiv.org/pdf/1201.4089.pdf. An excellent introduction to description logics describing the relationship with OWL2.

W3C Standards and Submissions

OWL Related

- XML Schema https://www.w3.org/XML/Schema

- RDF 1.1 Primer https://www.w3.org/TR/rdf11-primer/

- RDF 1.1 Turtle https://www.w3.org/TR/turtle/

- RDF 1.1 Concepts and Abstract Syntax https://www.w3.org/TR/rdf11-concepts/

- RDF Schema 1.1 https://www.w3.org/TR/rdf-schema/

- OWL 2 Web Ontology Language Quick Reference Guide (Second Edition) https://www.w3.org/TR/owl2-quick-reference/

- OWL 2 Web Ontology Language Document Overview (Second Edition) https://www.w3.org/TR/owl2-overview/

- OWL 2 Web Ontology Language: Structural Specification and Functional-Style Syntax (Second Edition) https://www.w3.org/TR/owl2-syntax/

- OWL 2 Web Ontology Language Manchester Syntax (Second Edition) https://www.w3.org/TR/owl2-manchester-syntax/

- OWL 2 Profiles: An Introduction to Lightweight Ontology Languages, by M. Krotzsch
  http://korrekt.org/papers/Kroetzsch_OWL_2_Profiles_Reasoning_Web_2012.pdf

Companion Standards and Submissions:

- SKOS Simple Knowledge Organization System Primer
  https://www.w3.org/TR/skos-primer/

- SKOS Simple Knowledge Organization System Namespace Document - HTML Variant https://www.w3.org/2009/08/skos-reference/skos.html

- Using OWL and SKOS
  https://www.w3.org/2006/07/SWD/SKOS/skos-and-owl/master.html

- SWRL: A Semantic Web Rule Language Combining OWL and RuleML
  https://www.w3.org/Submission/SWRL/

- SPIN: Overview and Motivation https://www.w3.org/Submission/spin-overview/

- SPARQL Query Language for RDF https://www.w3.org/TR/rdf-sparql-query/

- R2RML: RDB to RDF Mapping Language https://www.w3.org/TR/r2rml/

- Shapes Constraint Language https://www.w3.org/TR/shacl/

Conceptual Modeling and Ontology Engineering

Formal Ontology of Subject, by C. Welty and J. Jenkins, *Data & Knowledge Engineering* 31: 155–181, 1999.
    https://pdfs.semanticscholar.org/b93f/5ed4d27056f6a7517589b2f0c63530e1b4e0.pdf

Representing Classes As Property Values on the Semantic Web https://www.w3.org/TR/swbp-classes-as-values/

Talk: *Conceptual vs. Operational: A False Distinction?* by M. Uschold and D. McComb
    https://www.slideshare.net/UscholdM/conceptual-vs-operational-a-false-distinction

"The importance of distinguishing between terms and concepts" by M. Uschold
https://semanticarts.com/blog/terms-concepts-whats-important-pt-1/

"Finding and avoiding bugs in enterprise ontologies," by M. Uschold
http://ceur-ws.org/Vol-1586/codes1.pdf

Actual Ontologies

- **gist:** https://www.semanticarts.com/gist/

- Basic Formal Ontology (BFO): http://ifomis.uni-saarland.de/bfo/

- GoodRelations http://www.heppnetz.de/ontologies/goodrelations/v1

- Ontology for Media Resources 1.0 (by W3C) https://www.w3.org/TR/mediaont-10/

- Financial Industry Business Ontology (FIBO)
  https://www.edmcouncil.org/financialbusiness

- Talk: *Putting FIBO to Use: Some Brass Tacks* by M. Uschold
  https://www.slideshare.net/UscholdM/putting-fibo-to-use

Ontologies: What Are They and What Are They For?

- *Software Wasteland*, by D. McComb https://technicspub.com/software_wasteland/ Describes problems with the state of enterprise computing that ontology and semantic technology can play a key role in addressing.

- "A translation approach to formal ontologies," T. Gruber. *Knowledge Acquisition* 5(2): 199–200, 1993. http://ksl-web.stanford.edu/KSL_Abstracts/KSL-92-71.html. The original paper that kicked off the idea of using an ontology as a computational artifact.

- "Ontologies: principles, methods and applications," M. Uschold, M. Gruninger, *Knowledge Engineering Review* 11(2),1996. The first comprehensive introduction to the emerging field of ontologies and their applications.

- "Ontologies and semantics for seamless connectivity," M. Gruninger and M. Uschold. *SIGMOD Record*, 33(4):58–64, 2004. A short paper introducing the need and nature and uses of ontologies.

- Ontology and database schema: What's the difference?" Uschold, M. *Applied Ontology*, 10(3-4) pp. 243–258, 2015. Based on a talk given in 2011: https://www.slideshare.net/UscholdM/ontologies-and-db-schema-whats-the-difference.

A.4   ANSWERS TO EXERCISES

A.4.1   CHAPTER 1

Exercise 1: Draw a diagram highlighting the key concepts and relationships for the subject of registering corporations.

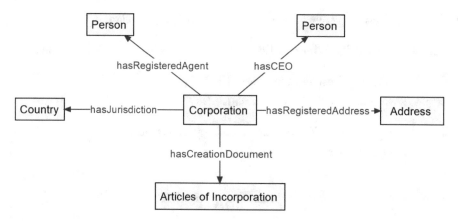

Exercise 2: How do you resolve the apparent contradiction that the U.S. Supreme Court's Citizen's United decision declared that a corporation is legally a person with the above common sense example in the ontology? Can you think about it in such a way that there is in fact no contradiction? **Answer:** The word "person" is being used ambiguously to mean two different things. One is a physical object and the other is a social object. Therefore, there is no contradiction.

A.4.2   CHAPTER 2

Exercise 1: First, see if you can determine which of the six kinds of assertions listed Table 2.3 are being made and which are not. Pay particular attention to whether, where and how object properties are used vs. data properties.

Answer

1. The four instances of `owl:Thing` are examples of assertions described in row 1 in the table.

2. The three links between the instances are object property assertion (row 5 in the table).

3. The nine green boxes inside the instances are data property assertions (row 6 in the table).

Exercise 2: Start with the statement "Every `PatientVisit` has at least one healthcare provider" and through a series of English statements, each rephrasing the prior until you end up with: "Every `PatientVisit` is the subject of at least one triple where the predicate is `careProvider` and the object is of type `Person`"

Answer

1. Every `PatientVisit` has at least one healthcare provider.

2. Every `PatientVisit` is associated with one `Person` who is providing care.

3. Every `PatientVisit` has a `careProvider` relationship with at least one `Person`.

4. Every `PatientVisit` is the subject of at least one triple where the predicate is `careProvider` and the object is of type `Person`.

Exercise 3: If you had to give that class "`doe:lastName some xsd:string`" a name, what might it be? What needs to be true to be a member of this class?

Answer: To be a member of the property restriction class "`doe:lastName some xsd:string`" an individual must have at least one last name that is a string. If you gave that class a name, "`ThingWithALastNameThatIsAString`" would be accurate.

Exercise 4: Can you think of other functional properties? Can someone have more than one first name? more than one last name? Can there be more than one person providing care on a given `PatientVisit`?

Answer: We saw that there can be multiple care providers on a single patient visit. If you count nicknames, people do have more than one first name. You might to say there is only one "official" first name, in which case the property could be functional. People can also use more than one last name. Some people choose to keep their own last name for most purposes, but use their spouse's last name for some official purposes. Again, there might be a single "official" last name that appears on your passport that could be functional. If your ontology is to be used for names in multiple countries, the notion of first and last name does not work so well. There are a lot of variations.

Other properties that are likely to be functional in most situations include the following:

- *identifies*: an identifier identifies no more than one thing.

- *hasDateOfBirth*: a person has only one date of date of birth

- *hasLatitude*: a place on the Earth has no more than one latitude. Other than the poles, each place only has one longitude as well. This is an edge case that you could choose to ignore and make `hasLongitude` functional as well.

- *hasVIN*: a vehicle only has one vehicle identification number

Exercise 5: Is the property `careRecipient` transitive? Why or why not?

Answer: It is not. The `careRecipient` always links a `PatientVisit` to a `Person`, it cannot be used to link a `Person` to anything. For a property to be transitive, the domain and range need to be the same.

Exercise 6: Explain why this works. How can just specifying the two-high level classes being disjoint make the others also disjoint?

Answer: Restated, the question is given that person and event are disjoint, how can we conclude that every class on the left side of Figure 2.15 is disjoint from every class on the right? Let's try to infer that cancer patient is disjoint from patient visit. We have to prove that those two classes cannot have any members in common. Let's see what happens if we assert an individual to be both a cancer patient and a patient visit.

The meaning of `rdfs:subClassOf` tells us that every cancer patient is also a patient, which, in turn, is also a person. It also tells us that every patient visit is an event. So our individual is inferred to be both a person and an event. But person and event are disjoint. So it is impossible for the individual to be both a cancer patient and a patient visit. Therefore, the two classes are disjoint.

A.4.4 CHAPTER 4

Exercise 1: Can you think of examples of real world properties that might be either a subproperty or superproperty of `isSubsidiaryOf`?

Answer:

Superproperty: `partOf`.
Subproperty: `whollyOwnedSubsidiary`

Exercise 2: What would the Venn diagram in Figure 4.3 have to look like to rule out being an employee?

Answer:

The `employedBy` and `contractorFor` ovals would have no overlap.

Exercise 3: Can you think of a good name for the inverse of `hasBorrower`? Try filling in the blank in this sentence:

"John Doe _____ loan contract 857466"

What about `worksFor`? Try filling in the blank:

"Apple _____ Tim Cook"

Answer:

`hasBorrower` "John Doe *is the borrower on* loan contract 857466." Thus, the inverse of `hasBorrower` could be `isBorrowerOn`.

`worksFor` "Apple *has the worker* Tim Cook." Thus the inverse of `worksFor` could be `has-Worker`.

NOTE: Sometimes there aren't any good names for inverse properties.

Exercise 4: Draw a Venn diagram like the one in Figure 4.9 for the property chain that goes from the care provider to the patient through the patient visit as depicted in Figure 4.11. Hint: the property chain you want is: `careProviderOn o careRecipient`. It will be a sub-property of `gaveCareTo`.

Answer:

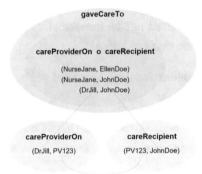

Exercise 5: Think of a few more properties in each category: symmetric, asymmetric and neither. Which categories were easy vs. hard to find properties for? Which category do you think will generally have the most properties? Why?

    Hint: What needs to be true about the domain and range of a symmetric property?

Answer:

Symmetic: is exchanged for, is topically related to, `owl:sameAs`

Asymmetric: employed by, care provider, has borrower

Neither: has favorite colleague, tends to result in, manages

When thing mutually tend to result in the other, you have either a vicious or virtuous cycle. One person could manage another person in one area, and be managed by that same person in another area.

A symmetric property must have the same domain and range. That is why there are relatively few. Most relationships occur between different kinds of things.

Exercise 6: Answer the following three additional questions asking what we can know about the characteristics of the inverse of a property from the characteristics of the property. Specifically:

1. If property p is functional, what can we say about whether p_inv is or is not functional?

2. If property p is symmetric, what can we say about whether p_inv is or is not symmetric?

3. If property p is transitive, what can we say about whether p_inv is or is not transitive?

Hint: it may help to draw some Venn diagrams representing properties as sets of pairs.

Answer:

1. If p is functional, p_inv need not be functional. A person has only one last name, but many people can have that last name.

   Similarly: If p is inverse functional, p_inv need not be inverse functional. This is for the same reason as the above.

2. If p is symmetric, p_inv is also symmetric. This follows from the fact that a symmetric property is the inverse of itself.

3. If p is transitive, p_inv is also transitive. This one is a bit less obvious. We can go through the following line of reasoning to convince ourselves. The transitivity of p means for every x, y, and z:

   IF: x p y and y p z
   THEN: x p z.

   Recall that asserting x p y is the same as asserting y p_inv x. So we can rewrite the above as:

   IF: y p_inv x and z p_inv y
   THEN: z p_inv x.

Since the order in the first line does not matter we can rewrite this as:

IF: z p_inv y and y p_inv x
THEN: z p_inv x.

This is the law of transitivity, but it is slightly hidden because the variables are not in the usual order. Let's make it more obvious. Since x, y, and z are totally arbitrary, we can use different variable names. If we substitute u for z, v for y, and w for x, we get:

IF: u p_inv v and v p_inv w
THEN: u p_inv w.

This is exactly what it means for p_inv to be transitive.

This sort of reasoning is sometimes needed to discover and convince yourself about the truth of some proposition. There is a way to see this more intuitively by thinking of an example. Let's say that the property partOf is transitive. A good name for the inverse is hasPart (see the top of Figure 4.20). The question is whether the bottom figure necessarily follows. After a while, I hope you can convince yourself.

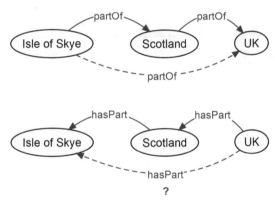

Figure 4.20: Transitivity of inverse.

Exercise 7: Can you think of another example where two properties will be disjoint? Hint: think about loans, borrowers and lenders.

Answer: If a loan contract is represented as an individual, the properties hasBorrower and hasLender could connect that contract to the two main parties. Since the borrower and lender cannot be the same party on a single loan, those two properties would be disjoint.

Exercise 8: A symmetric property is its own inverse. What similarly pithy thing can be said that characterizes the essence of what it means to be an asymmetric property?

Answer: An asymmetric property is disjoint from its inverse.

A.4.5 CHAPTER 5

Exercise 1: Create Venn diagrams to convince yourself that the inferences for intersection, union and complement are justified.

Answer:

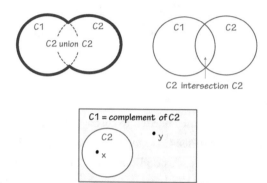

Exercise 2: Spell out the inferences regarding `TwoWheeledThing` in a more precise notation using triples.

Answer: The following expression in Manchester syntax,

```
Class: doe:TwoWheeledVehicle
       EquivalentTo:
 doe:hasPart exactly 2 doe:Wheel
```

sanctions the following inferences:

```
x rdf:type :TwoWheeledVehicle
```
IF AND ONLY IF

There exists two individuals `w1` and `w2` such that

```
w1 owl:differentFrom w2 and
w1 rdf:type :Wheel and
w2 rdf:type :Wheel and
x :hasPart w1 and
x :hasPart w2
```

Exercise 3: Can you convince yourself that asserting a given property, `p`, to be functional means the same thing as asserting that `owl:Thing` is a subclass of the restriction (`p max 1`)? Hint: draw a picture of a Venn diagram.

Answer: Note: this is not an easy exercise. It may take a while to understand the answer, even after reading it. To say that `owl:Thing` is a subclass of (`hasBiologicalMother max 1`) is to say that every member of `owl:Thing` (i.e., every individual, period) can have no more than one biological mother. But that is exactly what it means for `hasBiologicalMother` to be functional.

Exercise 4: Write down in English the meaning of each of the data property restrictions.

1. (`:firstName some xsd:string`)

   The set of individuals with at least one first name that is a string.

   The set of all individuals that are subjects of at least one triple using the predicate `firstName` whose object is a literal of datatype `xsd:string`.

2. (`:hasSSN all xsd:integer`)

   The set of all individuals that only have social security numbers that are integers.

   The set of all individuals that are subjects of triples using the predicate `hasSSN` whose objects are only literals of datatype `xsd:integer`.

3. (`:hasDescription min 1 xsd:string`)

   The set of all individuals that have at least one description that is a string.

   The set of all individuals that are subjects of at least one triple using the predicate `hasDescription`, whose object is a literal of datatype `xsd:string`.

4. (`:hasLicenseNumber max 1 xsd:string`)

   The set of all individuals that have at most one license number that is a string.

   The set of all individuals that are subjects of at most one triple using the predicate `hasLicenseNumber`, whose object is a literal of datatype `xsd:string`.

5. (`:hasCodeName exactly 2 xsd:string`)

   The set of all individuals that have exactly two code names.

   The set of all individuals that are subjects of exactly two triples using the predicate `hasCodeName`, whose objects are literals of datatype `xsd:string`.

6. (`:hasColor "red"^^xsd:string`)

   The set of all individuals that have color red, as a string

The set of all individuals that are subjects of triples using the predicate `hasColor`, whose object is the literal: `"red"^^xsd:string`.

A.4.6 CHAPTER 6

Exercise 1: Given our definition for a security agreement, what if all you know is the following:

```
:_x doe:partOf :_LoanContract_203 .

:_?b203 rdf:type :LoanContract .
```

What if any inferences can be made?

Answer: None. Being part of a loan contract is necessary but not sufficient to infer into the class, `SecurityAgreement`. The individual, `_x` might be one of several terms that are part of the contract. In general, the subclass usage of a restriction sanctions inferences if you know something is a member of a class, but you cannot infer into the class. For that you need to use equivalence.

Exercise 2: Draw a Venn diagram to make it easy to see why `:Commitment` is inferred to be a subclass of `:Intention`.

Answer:

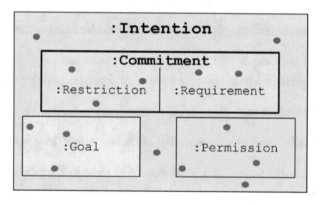

A.4.8 CHAPTER 8

Exercise 1: What justifies the inference of `Brother` into `Male` and `Sister` into `Female` in the examples shown in Figure 8.11.

Answer:

> **Top of figure:** A `Brother` must be a `brotherOf` some `Person`. The inverse of `brotherOf` is `hasBrother` which has range of `Male`. This means that the domain of `brotherOf` is `Male`. Thus a `Brother` has to be a `Male`.

Bottom of figure: The inference is easier in this case. A `Brother` is the intersection of two classes, one of which is `Male`. Therefore a `Brother` has to always be `Male`.

Author Biography

Michael Uschold has over 25 years of experience in developing and transitioning semantic technology from academia to industry. He pioneered the field of ontology engineering, co-authoring the first paper on the topic in 1995 that described his experiences as lead developer of the "Enterprise Ontology." He also co-authored the first comprehensive introduction to the emerging ontology field in 1996, a source still widely referenced to this day.

Michael is a world-class mentor and trainer. He co-presented the first ontology tutorial in 1995 in London and was an invited instructor at the first Semantic Web summer school in Madrid in 2003 as well as at the Second Interdisciplinary Summer School on Ontological Analysis in Vitoria, Brazil, in 2014. He has given numerous invited talks and tutorials at industry and academic conferences on semantic technology, emphasizing practical ways to put ontologies to use.

As a senior ontology consultant at Semantic Arts since October 2010, Michael trains and guides clients to better understand and leverage semantic technology. He has taught hundreds of budding ontologists about OWL in the past decade, in tutorials, open classes, and seminars held in-house in major organizations. He has built commercial enterprise ontologies in sports and entertainment, digital asset management, finance, healthcare, legal research, consumer products, electrical product specifications, manufacturing, corporation registration, and metadata management.

From 2008–2009, Uschold worked at Reinvent on a team that developed a semantic advertising platform that substantially increased revenue. As a research scientist at Boeing from 1997–2008, he defined, led, and participated in numerous projects applying semantic technology to enterprise challenges. He is a frequently invited speaker and panelist at national and international events and serves on the editorial board of the *Applied Ontology Journal*. He received his Ph.D. in Artificial Intelligence from Edinburgh University in 1991 and his M.Sc. in Computer Science from Rutgers University in 1982.

Index

Printed in the United States
by Baker & Taylor Publisher Services